CURRICULUM AND EVALUATION
S T A N D A R D S
FOR SCHOOL MATHEMATICS
ADDENDA SERIES, GRADES 9–12

GEOMETRY FROM MULTIPLE PERSPECTIVES

Arthur F. Coxford, Jr.

with

Linda Burks

Claudia Giamati

Joyce Jonik

Consultants

Harold L. Schoen

Daniel Teague

Christian R. Hirsch, Series Editor

NATIONAL COUNCIL OF
TEACHERS OF MATHEMATICS

Copyright © 1991 by
THE NATIONAL COUNCIL OF TEACHERS OF MATHEMATICS, INC.
1906 Association Drive, Reston, Virginia 22091-1593
All rights reserved

Fourth printing 1996

Library of Congress Cataloging-in-Publication Data:

Coxford, Arthur F.
     Geometry from multiple perspectives / Arthur F. Coxford, Jr., with
Linda Burks, Claudia Giamati, Joyce Jonik.
          p.   cm. — (Curriculum and evaluation standards for school
mathematics addenda series. Grades 9–12)
     Includes bibliographical references.
     ISBN 0-87353-330-5
     1. Geometry—Study and teaching (Secondary) I. Title.
II. Series.
QA461.C65   1991
516'.0071'2—dc20   91-16636
                                                                      CIP

The figures in figure 5.9 on page 37 originally appeared in the following publications:

Christie, Archibald H. *Pattern Design*. Mineola, N.Y.: Dover Publications, 1969.
Reprinted with permission of Dover Publications.

Crowe, Donald W. "The Geometry of African Art. II. A Catalog of Benin Patterns." *Historia Mathematica*
     2 (1975): 253–71.
Reprinted with permission of the Academic Press.

Dye, Daniel Sheets. *Chinese Lattice Designs*. Mineola, N.Y.: Dover Publications, 1974.
Reprinted with permission of Dover Publications.

Stevens, Peter S. *Handbook of Regular Patterns*. Cambridge, Mass.: MIT Press, 1980.
Copyright © 1981 by Massachusetts Institute of Technology, published by The MIT Press.

Williams, Geoffrey. *African Designs from Traditional Sources*. Mineola, N.Y.: Dover Pictorial Archive
     Series, 1971.
Reprinted with permission of Dover Publications.

Zaslavsky, Claudia. *Africa Counts*. Boston: Prindle, Weber, & Schmidt, 1973.
Reprinted with permission of the author.

Printed in the United States of America

TABLE OF CONTENTS

FOREWORD

As the *Curriculum and Evaluation Standards for School Mathematics* (NCTM 1989) was being developed, it became apparent that supporting publications would be needed to aid in interpreting and implementing the curriculum and evaluation standards and the underlying instructional themes. A Task Force on the Addenda to the *Curriculum and Evaluation Standards for School Mathematics,* chaired by Thomas Rowan and composed of Joan Duea, Christian Hirsch, Marie Jernigan, and Richard Lodholz, was appointed by Shirley Frye, then NCTM president, in the spring of 1988. The Task Force's recommendations on the scope and nature of the supporting publications were submitted in the fall of 1988 to the Educational Materials Committee, which subsequently framed the Addenda Project for NCTM Board approval.

Central to the Addenda Project was the formation of three writing teams to prepare a series of publications targeted at mathematics education in grades K–6, 5–8, and 9–12. The writing teams consisted of classroom teachers, mathematics supervisors, and university mathematics educators. The purpose of the series was to clarify the recommendations of selected standards and to illustrate how the standards could realistically be implemented in K–12 classrooms in North America.

The themes of problem solving, reasoning, communication, and connections have been woven throughout each volume in the series. The use of technological tools and the view of assessment as a means of guiding instruction are integral to the publications. The materials have been field tested by teachers to ensure that they reflect the realities of today's classrooms and to make them "teacher friendly."

We envision the Addenda Series being used as a resource by individuals as they begin to implement the recommendations of the *Curriculum and Evaluation Standards.* In addition, volumes in a particular series would be appropriate for in-service programs and for preservice courses in teacher education programs.

On behalf of the National Council of Teachers of Mathematics, I would like to thank the authors, consultants, and editor, who gave willingly of their time, effort, and expertise in developing these exemplary materials. Gratitude is also expressed to the following teachers who reviewed drafts of the material as this volume progressed: Kathryn Moore, Daren Starnes, and James Wheaton. Finally, the continuing technical assistance of Cynthia Rosso and the able production staff in Reston is gratefully acknowledged.

Bonnie H. Litwiller
Addenda Project Coordinator

The *Curriculum and Evaluation Standards for School Mathematics*, released in March 1989 by the National Council of Teachers of Mathematics, has focused national attention on a new set of goals and expectations for school mathematics. This visionary document provides a broad framework for what the mathematics curriculum in grades K–12 should include in terms of content priority and emphasis. It suggests not only what students should learn but also how that learning should occur and be evaluated.

Although the *Curriculum and Evaluation Standards* specifies the key components of a quality contemporary school mathematics program, it encourages local initiatives in realizing the vision set forth. In so doing, it offers school districts, mathematics departments, and classroom teachers new opportunities and challenges. The purpose of this volume, and others in the Addenda Series, is to provide instructional ideas and materials that will support implementation of the *Curriculum and Evaluation Standards* in local settings. It addresses in a very practical way the content, pedagogy, and pupil assessment dimensions of reshaping school mathematics.

Reshaping Content

The curriculum standards for grades 9–12 identify a common core of mathematical topics that *all* students should have the opportunity to learn. The need to prepare students for the workplace, for college, and for citizenship is reflected in a broad mathematical sciences curriculum. The traditional strands of algebra, functions, geometry, and trigonometry are balanced with topics from data analysis and statistics, probability, and discrete mathematics. Within the traditional strands, topics to receive decreased or increased attention are identified. In the case of geometry, the *Curriculum and Evaluation Standards* calls for a move away from geometry as an archives tour through an extensive collection of predetermined Euclidean theorems and their two-column proofs. It advocates greater attention to coordinate and transformation approaches, to real-world applications and modeling, and to investigations (often technology-based) leading to student-generated theorems with supporting arguments expressed orally or in paragraph form. A major goal is to build the kind of strong geometric intuition that has been shown to be an important factor for success on the job and in college.

Geometry from Multiple Perspectives links the content proposed in the *Curriculum and Evaluation Standards* to that of current programs. Ways for blending coordinate and transformational ideas with conventional synthetic ideas are described and supported by specific examples. Applications such as frieze patterns and fractals are included to add further interest to new and traditional topics and to highlight the usefulness of geometry in our world. Options for organizing the content of geometry more "locally" so as to emphasize reasoning as opposed to memorization are elaborated.

A special *Try This* feature appearing throughout the volume provides exercises, problems, and explorations for use with students. We hope that these will pique your interest and that you will use the margins productively. More extensive investigations and projects appear as blackline masters at the end of each chapter. Solutions and hints for these activities appear in the Appendix. Additional sources of ideas and materials supporting content themes in the *Curriculum and Evaluation*

Standards are identified in a selective annotated bibliography at the end of this volume.

Reshaping Pedagogy

The *Curriculum and Evaluation Standards* paints mathematics as an activity and process, not simply as a body of content to be mastered. Throughout, there is an emphasis on doing mathematics, recognizing connections, and valuing the enterprise. Hence, standards are presented for Mathematics as Problem Solving, Mathematics as Communication, Mathematics as Reasoning, and Mathematical Connections. The intent of these four standards is to frame a curriculum that ensures the development of broad mathematical power in addition to technical competence; that cultivates students' abilities to explore, conjecture, reason logically, formulate and solve problems, and communicate mathematically; and that fosters the development of self-confidence.

Realization of these process and affective goals will require, in many cases, new teaching-learning environments. The traditional view of the teacher as authority figure and dispenser of information must give way to that of the teacher as catalyst and facilitator of learning. To this end, the standards for grades 9–12 call for increased attention to—

- actively involving students in constructing and applying mathematical ideas;
- using problem solving as a means as well as a goal of instruction;
- promoting student interaction through the use of effective questioning techniques;
- using a variety of instructional formats—small cooperative groups, individual explorations, whole-class instruction, and projects;
- using calculators and computers as tools for learning and doing mathematics.

Geometry from Multiple Perspectives reflects the new methodologies supporting new curricular goals. The classroom-ready activity sheets at the end of the chapters provide tasks that require students to experiment, collect data, search for patterns, make conjectures, and give convincing arguments. These activities are ideally suited to cooperative group work. They often conclude with a challenge suitable for in-depth project work. The previously mentioned *Try This* feature appearing throughout the volume furnishes more structured tasks offering opportunities for genuine problem solving, reasoning, and lively classroom discourse.

Teaching Matters is another special feature of this book. These captioned margin notes supply helpful instructional suggestions, including ideas on motivation and on the use of appropriate concrete materials and technological tools. They also identify possible student misconceptions and some difficulties that students might encounter with certain topics and suggest how these can be anticipated and addressed in instruction.

At the heart of changing patterns of instruction are the growing potentialities of technology. The standards for grades 9–12 assume that students will have access to graphics calculators and that computers will be available for demonstration purposes as well as for individual and group work. Drawing and measuring computer utilities provide learning environments that foster the active process of making and exploring conjectures. How this software can be used to bring a new vitality and spirit of inquiry to school geometry is illustrated in chapters 3 and 4. The discussion is accompanied by several computer-based investigations in blackline master form. As with all student investigations, it is important that provisions be made for students to share their experiences, clarify their thinking, generalize their discoveries, construct convincing arguments, and recognize connections with other topics. Possible contributions of Logo turtle graphics and graphics calculators to learning and teaching geometry are also addressed in this volume.

Reshaping Assessment

Complete pictures of classrooms in which the *Curriculum and Evaluation Standards* is being implemented not only show changes in mathematical content and instructional practice but also reflect changes in the purpose and methods of student assessment. Classrooms where students are expected to be engaged in mathematical thinking and in constructing and reorganizing their own knowledge require adaptive teaching informed by observing and listening to students at work. Thus, informal, performance-based assessment methods are essential to the new vision of school mathematics.

Analysis of students' written work remains important. However, single-answer paper-and-pencil tests are often inadequate to assess the development of students' abilities to analyze and solve problems, make connections, reason mathematically, and communicate mathematically. Potentially richer sources of information include student-produced analyses of problem situations, solutions to problems, reports of investigations, and journal entries. Moreover, if calculator and computer technologies are now to be accepted as part of the environment in which students learn and do mathematics, these tools should also be available to students in most assessment situations.

Geometry from Multiple Perspectives reflects the multidimensional aspects of student assessment and the fact that assessment is integral to instruction. *Assessment Matters*, a special margin feature, provides suggestions for assessment techniques and for test items related to the content under discussion. Particular attention is given to alternative ways of assessing reasoning and proof-making.

Conclusion

The two standards on geometry for grades 9–12, informed by research and based on the wisdom of practice, represent the consensus of the profession on what the shape of school geometry today and tomorrow should be. Of all high school courses, geometry as it is commonly taught probably provides the fewest specific prerequisites for future study and work. As such, it offers a promising opportunity for initiating change by effectively blending new and traditional topics as described in this volume.

Sustainable change must occur first in the hearts, minds, and classrooms of teachers and then in their departments and school districts. Individually we can initiate the process of change; collectively we can make the vision of the *Curriculum and Evaluation Standards* a reality. We hope you will find the information in this book valuable in translating the vision of the *Standards* into practice.

Christian R. Hirsch, Editor
Grades 9–12 Addenda Series

CHAPTER 1
WHY SHOULD GEOMETRY BE CONSIDERED FROM MULTIPLE PERSPECTIVES?

In the world around us, we see manufactured objects. Many are made of parts that are linear or circular in shape. These parts are based on the geometry of Euclid—that is, they are based on segments, angles, triangles, quadrilaterals, polyhedra, circles, spheres, and so on. This is the geometry of the point set, of the straight line, and of the Euclidean tools of construction. Synthetic Euclidean geometry is appropriate for describing many aspects of our world. On the assembly line, we make congruent copies of gadgets, appliances, and vehicles. In offices and shops, we make scale models and blueprints based on this geometry. We construct road intersections perpendicular to each other so that drivers can see equally clearly in both directions. We locate the position to place a lighting fixture in the center of a ceiling by drawing diagonals of the rectangular model of the ceiling. These illustrations are only a small sample of the practical uses and explanatory power of Euclidean geometry, which has been hailed as a major human intellectual achievement and which has held a position of esteem in the school mathematics curriculum for many years.

But as we progress as a civilization, additional uses of geometric ideas arise both within and outside mathematics—uses that cannot be easily accommodated by the old representations of geometric entities. Certainly René Descartes recognized the power of using coordinates to represent points and the power of using algebraic statements to represent other geometric figures. Even though there is evidence that the Egyptians and the Greeks used coordinate ideas long before 1637, when *La Géométrie* was published, Descartes's contribution is significant. It marks the beginning of the powerful and useful integration of algebra and geometry into analytic geometry.

Today the interplay between algebra and geometry is even greater. On the one hand, in mathematics we use the language of geometry to describe concepts in abstract "spaces" such as hyperplanes and *n*-dimensional spheres. On the other hand, in plants, factories, and laboratories in the real world, computer-aided design uses coordinates and algebraic representations to describe geometric forms that are not linear or circular. For example, the cross section of an automobile fender may be described by a function of two variables once an appropriate coordinate system is established. With this function, the contours of the fender may be modified, and designers can "try out" their ideas for subtle changes without actually producing a physical model of the fender. Such capability gives the user the power to do things that were once only possible in the imaginations of engineers and architects. The idea is simple, yet powerful; namely, we can use numbers to identify locations, and we can change those locations according to some formula. The worker of tomorrow will need to understand this simple idea.

But all is not completely well in "geometry city." Concepts that were easily understood when described using synthetic techniques become cumbersome and less easily understood in the corresponding coordinate description. In synthetic geometry, a triangle is a union of line segments intersecting only at the endpoints. In the coordinate representation, we may need three linear equations defined over specific intervals in order to describe that same triangle. Sometimes the algebraic description is more useful, and at other times the synthetic description is more useful. This suggests that the two perspectives are complementary, not conflict-

◆　　　◆　　　◆　　　◆　　　◆　　　◆　　　◆　　　◆

ing. Both need to be recognized and understood in order to improve the power and problem-solving capacity of the student.

The development of computer technology has increased the importance of being able to represent shapes in either perspective and has added another—vectors. For example, a triangle may be defined by plotting commands similar to those of Applesoft BASIC:

```
10 PLOT 23,45 TO 49,45
20 PLOT TO 30,8
30 PLOT TO 23,45
```

Here the triangle is made up of the three vectors.

In Logo, a triangle may be defined as a procedure—TO TRIANGLE.

```
TO TRIANGLE
RT 90 FD 40
LT 135 FD 50
LT 135 FD 40
END
```

Drawing triangles or other polygons with computer graphics may require the student to use coordinates, vectors, or unusual synthetic aspects of the figure. In the Logo procedure above, the student needs to recognize that the 135-degree angles are exterior angles of the triangle. Further information on the uses of Logo may be found in Kenney (1987).

Yet examining geometry from both the synthetic and the coordinate perspectives is not sufficiently rich to accommodate the multiple uses of geometry today. Earlier we suggested that the assembly line, on which congruent copies are produced, is an example of the importance of synthetic geometry. That illustration is incomplete in that the congruence of Euclid applies only to circles and to rectilinear shapes common to the school curriculum. That is, this notion of congruence is essentially tied to the notion of angles, segments, and triangles. Fortunately, the objects we manufacture on the assembly line are not limited to being rectilinear objects and their unions. We demand that two fenders coming off the line must be congruent in order for them to be useful in constructing an automobile. This suggests that we need to expand our notion of congruence to include geometric shapes in addition to segments, angles, triangles, and circles. Curves can be congruent, as well as surfaces and solids in three-dimensional space.

The same argument holds for the notion of similarity. It is not sufficient for similarity to be applicable only to segments, to angles, to triangles, and to circles. Scale models of Corvettes are similar to the real thing, yet there is scarcely a triangle to be seen, and the "lines" of the car are not the "line segments" of synthetic Euclidean geometry. Thus we need to be able to extend the notions of congruence and similarity to shapes that are more complex than the usual Euclidean fare. But we may also wish to extend the ideas of congruence and similarity to sets of points that are less complex as well, such as to finite sets of points. Consider, for example, the point sets in figure 1.1.

The idea that is needed here is that of geometric transformation. The applicable geometric transformations are the isometries—reflection, rotation, translation, and glide reflection—and the similarities. These transformations are functions that have as their domains the sets of points in the plane or space and that have as their ranges the same

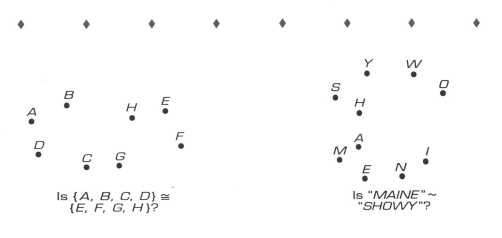

Is {A, B, C, D} ≅ {E, F, G, H}?

Is "MAINE" ~ "SHOWY"?

Fig. 1.1

sets of points. In practice, however, the user is usually concerned only with a shape and its image rather than with the entire plane or space.

The strengths of using transformations to define congruent figures and similar figures are easily enumerated. First, since the domain is a set of points, there are no restrictions on the nature of that set. It can be a finite set as in the illustration above, or it can be a Euclidean shape such as a square or a triangle, or it can be a "complex" figure or a curve that describes the cross section of a fender for a Corvette. Second, since the transformations needed are functions, the study of geometry is connected more closely to the study of other branches of mathematics in which function plays a central and more explicit role. Third, since transformations can be thought of as models of the mental or physical manipulations we use on shapes to determine if figures are congruent or similar, the mathematical ideas are more closely connected to the experience of the learner. Most of us have seen a youngster take a model of a figure, manipulate it to see if it will fit exactly on another figure, and in so doing, determine that the two figures are congruent. Fourth, we may synthetically represent the transformations themselves by using coordinate methods or matrices. By using matrix notation to represent the coordinates, students are able to experience the extraordinary power of graphics calculators with matrix capabilities. This emphasizes further the connections among the branches of mathematics and helps students attain the goals associated with the "mathematical connections" standard of the *Curriculum and Evaluation Standards for School Mathematics* (NCTM 1989).

The availability of transformation methods increases the range of problems accessible to the neophyte problem solver. As with the coordinate and synthetic perspectives, some situations are more easily represented and analyzed using transformations than they are with other tools.

For example, in figure 1.2 consider the question of where a power transformer should be located on $\overleftrightarrow{AB}$ so that the length of the cable needed to run to C and to D from $\overleftrightarrow{AB}$ is minimized. We note first that if C and D are on opposite sides of $\overleftrightarrow{AB}$, then the segment joining C and D will intersect $\overleftrightarrow{AB}$ in the appropriate position for the transformer. But in the example illustrated, the locations C and D are on the same side of $\overleftrightarrow{AB}$. The solution is to reflect either D or C over $\overleftrightarrow{AB}$ and to join the image to the remaining unreflected point. Since the straight line is the shortest distance and distance is preserved under reflection, the intersection of $\overleftrightarrow{AB}$ and the drawn line is the desired location.

To this point we have given arguments for approaching school geometry from multiple perspectives that emphasize today's needs. There is also an argument that addresses the needs of the future. It dates from 1975

Teaching Matters: Cooperative group work furnishes an opportunity for developing communication skills and for conducting informal, individual assessments that are not easily obtained during direct instruction or individual seatwork. As groups work on problems such as the transformer problem here, circulate around the room, unobtrusively observing the level of participation of individual students and how well the students apply key concepts and problem-solving strategies. Have a member of several different groups share their progress or solutions with the rest of the class. Demonstrating respect for students' ideas fosters the development of their mathematical power.

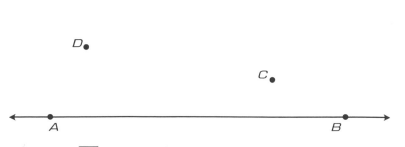

Where on $\overline{AB}$ should a transformer be placed so that the distance from D to the transformer to C is a minimum?

Fig. 1.2

and lies in the development of fractal geometry. The geometry of Euclid, the introduction of coordinates and algebraic representations, and the use of transformation approaches still model only a small fraction of the objects naturally occurring in nature. Until the discovery of fractals in 1975 by Benoit B. Mandelbrot, objects such as the coast of Norway, or the frond of a fern, or a mountain range of Alaska defied geometric description. They were simply too irregular to be modeled using the usual geometric/algebraic tools. But with the development of computers, computer graphics, and fractals, these natural objects, with all their irregularities, can now also be modeled geometrically. Many of these models are based on the notion of self-similar sets. This means that wherever we look at a set, under whatever magnification, what we observe appears to come from the original set before magnification.

Such is the paradox of irregularity. Some figures are, at once, so irregular as to defy accurate description and so regular as to be called self-similar. In a real sense, the problem is that they are actually too regular for accurate description with the traditional tools of geometry. It is in the area of fractal geometry that the language and notation of algebra and function are most powerful. The process of describing and creating self-similar objects is described mathematically through the iteration of functions, which can be thought of in a simple way as repeated compositions of a function with itself. Geometry and algebra meet again but in a very different context from that foreseen by Descartes.

Geometry, today and tomorrow, must be approached from multiple perspectives to permit the user to make the most of the content as its uses broaden and expand into heretofore unknown regions of science and nature. Fractals, which are founded on the concept of similarity, which are represented graphically (visually), and which are a creation at least partially dependent on powerful computers for their existence, are the new geometric tool of the near future. But what of the more distant future? What will the new tool be? What geometric content will it build on? No one knows, but you can be sure that it will demand an awareness of geometry from multiple perspectives for its comprehension.

In the rationale above, we have indicated that geometric ideas should, as recommended in the *Curriculum and Evaluation Standards* (NCTM 1989), be approached from a variety of perspectives: synthetic, coordinate, transformation, and vector. In what follows, we shall illustrate how these themes may be incorporated into the school geometry course as it is presently taught, by relating each theme to commonly taught content such as triangles, quadrilaterals, congruence, and similarity. In addition, we shall make suggestions for implementing other curriculum standards. In order to carry out this agenda, we shall begin with a brief review of some of the fundamental ideas needed, and then we shall turn in earnest to multiple perspectives of geometry.

CHAPTER 2
ELEMENTS OF CONTEMPORARY GEOMETRY

In the previous chapter, we presented a rationale for using coordinate and transformational ideas in school geometry as well as the usual synthetic emphases. In each of these approaches, attention is centered on sets of points. Geometric figures such as segments, lines, angles, polygons, polyhedra, spheres, cones, and planes are each sets of points that are subsets of the universal set called space. In synthetic geometry we draw these figures in an unreferenced space or plane; in coordinate geometry we add a reference system and "coordinatize" important points of the figures; in transformational geometry we move figures around following specific rules in either a coordinatized or an uncoordinatized environment. In all perspectives we seek to discover patterns among figures or within a fixed figure.

FUNDAMENTAL IDEAS

In synthetic geometry the key ideas are intersection; the measures of length, angle, and region; perpendicularity; parallelism; congruence; and similarity. Congruence of segments and angles is defined in terms of measures. Congruence and similarity are studied in detail for triangles. Usually these relations are defined as correspondences between vertices such that corresponding parts are all congruent or, for similarity, proportional and congruent. Several well-known criteria are presented from which one can conclude the congruence or the similarity of two triangles. But what about cases involving pairs of quadrilaterals or other polygons?

Are two quadrilaterals congruent if they satisfy SAS = SAS? ASAS = ASAS? SASAS = SASAS? SASAA = SASAA?

Coordinates are introduced in the usual course by means of the number line and the assumption that the points on the line are in one-to-one correspondence with the real numbers. This is a good place to introduce the distance between A, with coordinate a, and B, with coordinate b, as $|a - b|$ or $|b - a|$ (fig. 2.1).

$$AB = |a - b| = |b - a| = BA$$

Fig. 2.1

At this juncture the coordinate plane can also be introduced, and distance on the x- and y-axes can be developed. Later this idea can be extended to lines parallel to either axis and ultimately to the entire plane when the Pythagorean theorem is reached.

Distance is a key tool used in the coordinate perspe[...]
determine the congruence of segments. Two other [...]
in the coordinate perspective. They are midpoints (o[...]
points dividing a line in a specific ratio) and equations[...]
topics have been studied in algebra, but they should[...]
reviewed, or reinterpreted so that they will be availab[...]
needed in the coordinate perspective.

The fundamental ideas in the transformation perspec[...]
formations themselves, namely, reflection over a line[...]

Try this: *Give students a figure (for example, an isosceles triangle) and a coordinate grid. Ask them to give the coordinates of the vertices when the figure is placed on the grid. What placements of the figure produce the "best" coordinates? Explain what is meant by "best." Repeat for other figures known to the students.*

Teaching Matters: *An excellent way to investigate conditions of congruence is to give students particular measurements of angles and sides and have them draw and cut out models satisfying the conditions. These models can be placed on an overhead projector for comparison purposes. If the shapes are not congruent, the condition is not sufficient. If the shapes are congruent, then the condition is plausible, but it will need to be justified before it can be accepted.*

tion about a point, and dilation; composition of transformations; and the preservation properties of each transformation. A definition of each transformation is given below, along with an illustrative diagram.

Reflection: A' is the reflection image of A over the mirror line m, iff m perpendicularly bisects segment AA' (fig. 2.2). We write $r_m(A) = A'$, or just $r(A) = A'$ when no confusion will arise. Shown also is the reflection of triangle ABC over m. $\triangle A'B'C' = r_m(\triangle ABC)$ iff $A' = r_m(A)$, $B' = r_m(B)$, and $C' = r_m(C)$.

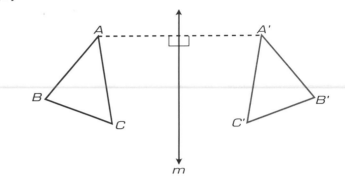

Fig. 2.2. Reflection

Miras and tracing paper can be used to investigate reflections and their properties in the physical world. By analyzing the relationship between points and their images relative to a "reflecting line," students can formulate their own mathematical description of a reflection. The computer software package Geometric Connectors: Transformations gives a computer-based means to investigate reflection.

The composition of two reflections is performed by reflecting successively over two mirror lines. Translation is defined as a composite of two reflections over parallel lines.

Translation: Figure F' is the translation image of figure F iff $r_n \circ r_m (F) = F'$, where $m \parallel n$. We write $T(F) = F'$. In figure 2.3, $\triangle ABC$ is the figure and is first reflected over line m, giving $\triangle A^*B^*C^*$. Then $\triangle A^*B^*C^*$ is reflected over line n. The result is $\triangle A'B'C'$. When these two reflections are composed ($r_n \circ r_m$), the translation, T, maps $\triangle ABC$ onto $\triangle A'B'C'$: $T(\triangle ABC) = \triangle A'B'C'$. The magnitude of the translation shown is the length of the segment AA'; the direction is along the perpendicular from the first line to the second. $T(\triangle ABC) = \triangle A'B'C'$ iff $T(A) = A'$, $T(B) = B'$, and $T(C) = C'$. Translations provide a simple way to introduce the idea of a vector. The directed distance that $\triangle ABC$ slides is the vector **AA'**.

Assessment Matters: Some errors on homework or test exercises will be indicative of difficulty with functional notation and not of basic misconceptions about transformations. Probing questions can be used to identify the nature of the student's difficulty so that the teacher can decide on appropriate corrective action.

Try this: Investigate how $r_m \circ r_n$ and $r_n \circ r_m$ differ when lines n and m are parallel.

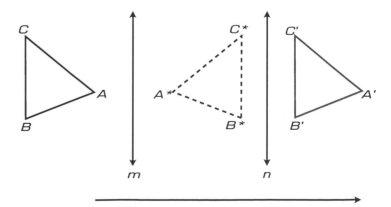

Fig. 2.3

A translation can be represented in the physical world with manipulatives. Any time an object is slid from one point to another without any twisting or flipping of the object, a real-world translation has occurred. Students should try to accomplish the same translation by reflecting successively over parallel lines.

Rotation: Figure F' is the rotation image of figure F iff $r_n \circ r_m (F) = F'$, where m and n intersect at a point, O (fig. 2.4). We write $R(F) = F'$. The magnitude of the rotation is $m < COC'$. The direction is from ray OC to ray OC'; in this example, clockwise. $R(\triangle ABC) = \triangle A'B'C'$ iff $R(A) = A'$, $R(B) = B'$, and $R(C) = C'$.

The physical representation of a rotation is that of a turning motion or movement along the circumference of a circle. A good way to illustrate this is to use tracing paper. Place the tracing paper over the figure, trace it, place your pencil point on the center of the rotation, and turn the tracing paper. As with translation, no flipping occurs, but the definition uses reflections. It would be worthwhile to discuss the value of this definition mathematically. This choice allows the relationship among the transformations to be easily seen, even though it is not intuitive in the physical representation.

Another way to illustrate rotations is to use the computer software package Geometric Connectors: Transformations. In this environment, the circular arcs are shown as the rotation occurs.

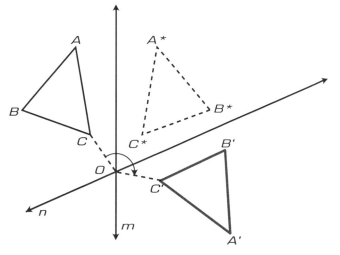

Fig. 2.4. Rotation

Dilation: A' is the image of A under a dilation with center O and magnitude r iff A' lies on ray OA and $OA' = r \cdot OA$ (fig. 2.5). We write $S(A) = A'$. $S(\triangle ABC) = \triangle A'B'C'$ iff $S(A) = A'$, $S(B) = B'$, and $S(C) = C'$.

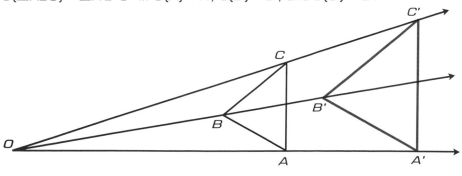

Fig. 2.5. Dilation

You can illustrate the idea of dilation with a copy machine that will enlarge or reduce. Have each student make a figure, reduce or enlarge it,
and then reduce or enlarge that copy by the same percentage. Cut out
the three figures. Place them on a sheet of paper so that the corresponding vertices are aligned and fall on a line (fig. 2.6). Compare the
ratios of the distances from the center of the dilation to the vertices. If
the reduction was 74 percent, what role does the factor 0.74 play?

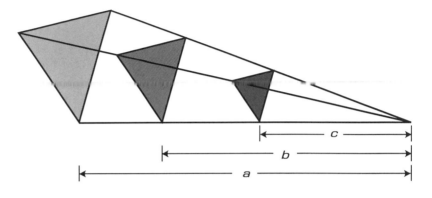

Fig. 2.6

Try this: Given that $r_m(A) = A'$
and $r_m(B) = B'$, prove that
$AB = A'B'$. That is, prove that
reflection preserves distance.

Reflections and composites of reflections preserve distance. They are
called isometries for this reason. Dilations preserve ratios of distances.
In figure 2.5, $A'C'/AC = B'C'/BC = r$. Composition, reflections, and
dilations give us the tools we need for all the congruence and similarity
theory in school geometry.

Try this: Show that $A'C' =$
$r \cdot AC$ in figure 2.5. That is,
prove that dilation preserves
ratios of distances.

REPRESENTATIONS

Representing figures and relations in synthetic geometry is familiar to
all high school mathematics teachers. We draw figures, label them with
capital letters, construct bisectors, add auxiliary lines, and reason about
the relationships represented in the figures.

When a coordinate perspective is introduced, other representations
are needed. Because polygons in the plane and polyhedra in space are
completely determined by their vertices, the usual synthetic sketches
are augmented by assigning coordinates to their vertices. Thus the
coordinate perspective of an isosceles triangle could be represented in
the plane by figure 2.7.

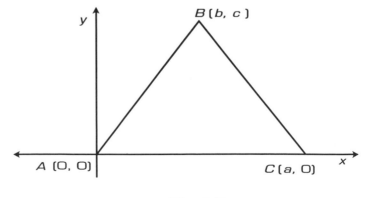

Fig. 2.7

But this representation does not take into account the properties unique to the isosceles triangle, for $B(b, c)$ is an appropriate representation for the third vertex of *any* triangle. The congruence of sides AB and BC can be used to improve the assignment of coordinates. First we note that $a > b$. Then

$$AB = BC \Leftrightarrow \sqrt{b^2 + c^2} = \sqrt{(a - b)^2 + (0 - c)^2}$$
$$\Rightarrow b^2 + c^2 = (a - b)^2 + c^2$$
$$\Rightarrow b^2 = (a - b)^2$$
$$\Rightarrow b = a - b, \text{ since } a > b$$
$$\Rightarrow b = \frac{a}{2}.$$

Thus better coordinates for B are $(a/2, c)$.

Above we have implied that distance (length) and the distance formula are essential for making the ideas of segment congruence useful in the coordinate perspective. Lines in this perspective are represented by the linear equation studied extensively in algebra. Lines in the plane are either parallel or intersecting. If parallel, the slopes of the equations are identical. Thus parallelism is reduced to the examination of two numbers. Since such examination does not distinguish between distinct and concurrent lines, the synthetic definition of parallel may need to be extended to include concurrent lines.

Similarly, when the lines intersect, the common coordinates are the solution to the system of linear equations representing the lines. Thus the solution of linear systems needs to be revived from its introduction in algebra. It should be noted also that when the coordinates of two points are identical, the two points are one. This is especially crucial when reasoning about polygons and intersecting diagonals, medians, altitudes, and angle bisectors. These lines often intersect in points that have a simple placement on a line, such as a midpoint or a trisection point. Geometric Connectors: Coordinates is a software package that allows you to draw and investigate most common geometric figures constructed by connecting points in a coordinate plane. In this environment, you have the advantages of being able to find coordinates of points and equations of lines, thus being able to profit from the strengths of two representations.

The ordered-pair representation of a point may also be thought of as a representation of a vector. This vector begins at the origin and ends at the point. It is called the position vector. It is equivalent to any vector for which the differences of the *x*- and *y*-coordinates of the tip and beginning points give the ordered pair of the position vector.

The coordinate perspective also permits the use of matrices to represent figures. Usually the point (or vector) $A(a, b)$ is represented by the column matrix $\begin{bmatrix} a \\ b \end{bmatrix}$, where the *x*-coordinate is in row 1 column 1, and the *y*-coordinate is in row 2 column 1. This matrix is also called a column vector.

A polygon is determined by its vertices. Thus a matrix can be used to represent the polygon by collecting all the column matrices of the vertices into one matrix. For the isosceles triangle in figure 2.7,

$$\triangle ABC = \begin{bmatrix} 0 & \frac{a}{2} & a \\ 0 & c & 0 \end{bmatrix}$$

Try this: **Find coordinates for the vertices of a parallelogram, a rhombus, a rectangle, an isosceles trapezoid, and a square with one vertex at the origin and a side along the x-axis.**

Try this: **Where should the coordinate system be placed to simplify the coordinate selection for a kite? A rhombus? A square?**

Teaching Matters: **The Curriculum and Evaluation Standards *recommends* that graphics calculators be available to all students at all times. In geometry, a graphics calculator can be used to represent polygons. Have students plot the coordinates of a parallelogram with sides of lengths 5 and 3, for example. Use the segment command to complete the figure. Do the same for other polygons of your choice.**

in which columns 1, 2, and 3 are the column vectors of A, B, and C. We shall find the matrix representation useful when transformations are introduced.

Transformations needed in Euclidean geometry were defined in the previous section. Figures are congruent if they can be mapped one onto the other by a composite of reflections. Figures are similar if they can be mapped one onto the other by a composite of reflections and dilations. Synthetically, these definitions give us a powerful tool for determining congruence and similarity of all figures, not just triangles, circles, segments, and angles. For those transformations that keep the origin fixed, the composites can be represented as products of matrices.

In the synthetic perspective the transformations are represented by drawings and are symbolized as follows:

$r_m(X)$ is   the reflection of X over the line m.
$T_v(X)$ is   the translation of X in the direction of vector $\mathbf{v}$.
$R_\phi(X)$ is   the rotation of X through an angle with measure ϕ.
$S_r(X)$ is   the dilation image of X, magnitude r.

When coordinates are added, the transformations may have quite convenient matrix representations. These representations allow images to be found by multiplying matrices or by adding column matrices. Since the vectors (1, 0) and (0, 1) form a basis for the plane, finding their images under a transformation gives the two-by-two matrix representing the transformation. The procedure follows.

1. Find the image of $\begin{bmatrix} 1 \\ 0 \end{bmatrix}$, say $\begin{bmatrix} a \\ b \end{bmatrix}$.

2. Find the image of $\begin{bmatrix} 0 \\ 1 \end{bmatrix}$, say $\begin{bmatrix} c \\ d \end{bmatrix}$.

3. Construct the transformation matrix by placing these images into the columns, namely, $\begin{bmatrix} a & c \\ b & d \end{bmatrix}$.

As an illustration, let us say we wish the matrix representation of reflection over the x-axis (fig. 2.8).

1. The image of (1, 0) is (1, 0).
2. The image of (0, 1) is (0, −1).
3. The matrix representation of reflection over the x-axis is $\begin{bmatrix} 1 & 0 \\ 0 & -1 \end{bmatrix}$.

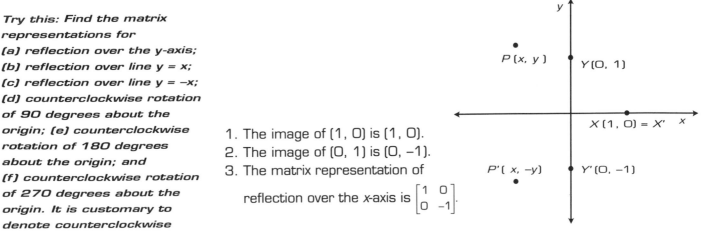

Fig. 2.8

Note that $r_{x\text{-axis}}(P) = P'$ where $P'(x, -y)$. To get this result using matrix techniques, multiply the column vector of (x, y) by the transformation matrix as shown.

$$\begin{bmatrix} 1 & 0 \\ 0 & -1 \end{bmatrix} \cdot \begin{bmatrix} x \\ y \end{bmatrix} = \begin{bmatrix} 1 \cdot x + 0 \cdot y \\ 0 \cdot x + -1 \cdot y \end{bmatrix} = \begin{bmatrix} x \\ -y \end{bmatrix}$$

This is the same result that was obtained by applying the properties of the reflection to $P(x, y)$.

Both the Sharp EL-5200 Super Scientific Calculator and the Texas Instruments TI-81 Graphing Calculator offer matrix operations in addition to point- and segment-plotting capabilities. With these calculators, students can perform all these transformations on complicated figures with relative ease.

Dilations centered at the origin are also easily represented using matrices. Recall that the magnitude is the key player here. Let r be the magnitude of a dilation with center at the origin.

By the definition, $S_r(A) = A'$ implies that $OA' = r \cdot OA$. But when A is $(1, 0)$ (fig. 2.9), $OA = 1$ and $r \cdot OA = r$. Thus A' is $(r, 0)$. Similarly for $B(0, 1)$, the image B' is $(0, r)$. Thus the matrix representing S_r is $\begin{bmatrix} r & 0 \\ 0 & r \end{bmatrix}$.

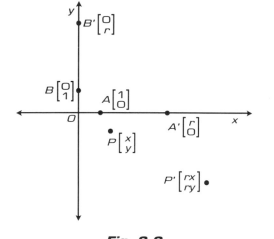

Fig. 2.9

To apply S_r to a point P with coordinates (x, y), we do the following:

$$P' = \begin{bmatrix} r & 0 \\ 0 & r \end{bmatrix} \cdot \begin{bmatrix} x \\ y \end{bmatrix} = \begin{bmatrix} r \cdot x + 0 \cdot y \\ 0 \cdot x + r \cdot y \end{bmatrix} = \begin{bmatrix} rx \\ ry \end{bmatrix} = (rx, ry)$$

Finally, translations can be represented by a vector that is made up of the horizontal and vertical components of the translation. For example, a translation that takes $(5, 7)$ onto $(8, -2)$ is given by **(3, –9)**. Thus to get the image of any point under $T_{(3, -9)}$, simply add **(3, –9)** to each point (x, y): $T(x, y) = (x, y) + (3, -9) = (x + 3, y - 9)$. If $(x, y) = (5, 7)$, as above, $T_{(3, -9)}(5, 7) = (5, 7) + (3, -9) = (8, -2)$. To get the translation image of any set of points, simply add the vector to the coordinates of each point.

In the discussion above, we have tried to outline some of the key ideas that will be used in viewing geometry from multiple perspectives. In the following chapters, these ideas will be found in specific illustrations, which could be introduced into your geometry instructional plan.

Try this: Given the pentagon ABCDE with vertices A(–3, 0), B(–1, –2), C(1, –2), D(3, 0), and E(0, 2), enter the matrix form on your calculator. Use the transformation matrices derived previously to find the coordinates of the transformed pentagon.

Try this: Choose three noncollinear points in a coordinatized plane. Join them to make △ABC. Represent △ABC in its matrix form. Find the image of △ABC under S_3 and $S_{0.5}$. Make the sketches. Do the same for octagon ABCDEFGH where A(–10, 3), B(–8, 0), C(–6, 0), D(–5, 1), E(5, 1), F(6, 0), G(8, 0), and H(10, 3).

Assessment Matters: When analyzing student performance on problems like the above, be sure to distinguish among errors in the following steps: (a) representing △ABC as a matrix, (b) computing the matrix product, and (c) making the sketch. Furthermore, in each step, an error may be a minor "slip" or it may be an indicator of a major misconception.

Initially triangles are defined, illustrated, and classified by the characteristics of their sides (scalene, isosceles, equilateral), of their angles (acute, right, obtuse), or of both (scalene-obtuse). A good activity to emphasize understanding and reasoning is to ask students to comment on multiple classifications of triangles. For example, if a triangle is drawn randomly, could it be (1) scalene and obtuse, (2) isosceles and obtuse, or (3) equilateral and obtuse? Similar pairings of categories may be made, and in each instance, the students must support their responses with an oral argument or a drawing illustrating a triangle with both characteristics. This activity provides experience with another valuable skill, namely, sketching shapes that are described verbally.

Computer drawing and measuring software is useful for investigating triangles and their properties. These software packages can be used to draw figures; to measure angles, segments, perimeters, and areas; and to construct such lines as angle bisectors, medians, and altitudes. These software packages include the Geometric Supposer programs (Apple II and IBM), Geometry One (IBM), Geometric Connectors: Coordinates (Apple II), and The Geometer's Sketchpad (Macintosh family). (A review of the Supposer and Geometry One software can be found in the fall 1989 [number 31] issue of *Consortium,* the newsletter of the Consortium for Mathematics and Its Applications [COMAP].)

Drawing and measuring software packages can be used to investigate characteristics of triangles. For example: At what point interior to an equilateral triangle is the sum of the distances from the point to the sides the greatest? Students could use the software to draw an equilateral triangle and to find the distances from several interior points to the sides. An examination of the data in small groups, individually, or as a class should lead to the observation that the sum of the lengths seems to be constant, that is, it seems to be independent of the point chosen. This is a surprising conjecture, but it can be proved easily by considering the area of the equilateral triangle and the areas of the three constituent triangles with the altitudes as the distances. In figure 3.1 we see that $\frac{1}{2}a \cdot s + \frac{1}{2}b \cdot s + \frac{1}{2}c \cdot s = \frac{1}{2}hs$. Thus $a + b + c = h$ and is constant.

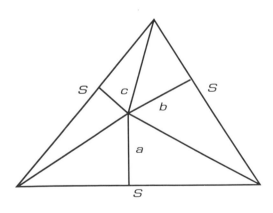

Fig. 3.1

The triangle has one special characteristic that is shared by no other plane shape and that makes the triangle vital in industry. An examination of the bridge in figure 3.2 shows that triangles form the basic construc-

Try this: Ask each student in the class to make a careful sketch of a triangle. Now have each student measure his or her triangle and categorize it by the nature of its sides and its angles. Collect this information in tabular form on the chalkboard or the overhead projector. Discuss the results. What triangle type is most commonly drawn? Least commonly drawn?

Assessment Matters: If you use a drawing and measuring utility in class, include some test items that either require students to use the utility or present some results as if the utility had generated them. Also assess each student informally while he or she explores with the utility. It is important that assessment techniques match instructional goals as closely as possible.

tion unit. Similarly, prefabricated roof trusses (fig. 3.3) use triangles extensively. The triangle is used because it is a rigid shape. If its vertices were fastened so that they could pivot, the triangle would retain its shape. You might ask students to explain the "extra" triangles used in the truss and the purpose of the vertical segments.

Try this: **Use a drawing and measuring utility to investigate the relation between the bisectors of an interior angle of a triangle and its corresponding exterior angle.**

Reprinted with permission from *Geometry in Our World*, copyright 1987 by the National Council of Teachers of Mathematics.

Fig. 3.2

Try this: **Have students use paper fasteners and tagboard strips with a hole punched in each end to investigate the rigidity of various polygon shapes. For shapes that are not rigid, determine how they can be made so.**

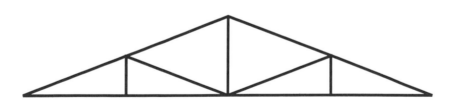

Fig. 3.3

When introducing triangles, place them on a coordinate plane once the defining characteristics are relatively clear to students. Triangles can be drawn and numerical coordinates can be assigned to vertices. As soon as the midpoint formula is reintroduced, apply it to the sides of triangles in the coordinate plane. Also give lines, in equation form, and have students draw the triangles determined. Have the students assign coordinates to the vertices by solving the systems of linear equations. Variable coordinates need to be used also, but extensive use should be postponed until slope and the distance formula are introduced.

Try this: **Have students make paper right triangles and find the midpoint of the hypotenuse. How is the midpoint related to the vertices of the triangle? Can they justify their conjectures?**

Later in the usual instructional sequence, medians and altitudes for triangles are introduced. Coordinate representations are useful for these topics. For example, suppose you are given the isosceles triangle ABC with $A(2, 0)$, $B(0, 2)$, $C(5, 5)$, and C is the vertex angle. Show that the median from C to $\overline{AB}$ is the altitude from C to $\overline{AB}$. The argument is easy. The midpoint of $\overline{AB}$ is $M(1, 1)$. Thus the slope of $\overline{MC}$ is 1. But the

◆   ◆   ◆   ◆   ◆   ◆   ◆   ◆

Try this: Given △ABC with A(2, 3), B(5, 5), and C(−1, 2), find the equations of the altitudes. Find the intersection of any two altitudes. Is the point on the third altitude? Do the same with medians.

Try this: Find a full-face picture of a person that appears to have bilateral symmetry. Make three overhead transparencies and cut two of them along the symmetry line. On the overhead projector, show the two left halves as a right and left face, the two rights, and the full face. Note the differences in the appearances of the three faces.

slope of $\overline{AB}$ is −1. Thus $\overline{MC}$ is the altitude as well as the median, since there is only one perpendicular from C to $\overline{AB}$. A similar argument may be used for variable coordinates.

Transformations are also useful in the study of triangles. Two transformations are particularly important—reflection and rotation. Reflection is the basis for *bilateral* or *reflection* symmetry. Rotation is the basis for *turning* or *rotational* symmetry. See figure 3.4.

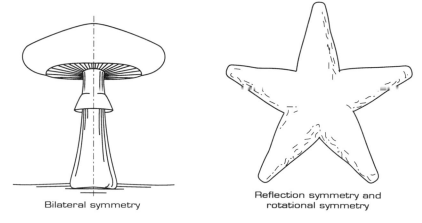

Bilateral symmetry

Reflection symmetry and rotational symmetry

Fig. 3.4

Triangles may have bilateral and rotational symmetry. For example, the isosceles triangle (fig. 3.5) has bilateral symmetry with respect to the line that is the angle bisector of the vertex angle. In the language of reflections, we say that △ABC is its own image under the reflection over line *CM*. We write

$$r(\triangle ABC) = \triangle BAC.$$

Thus we can conclude that

$$r(A) = B, \; r(B) = A, \text{ and } r(C) = C.$$

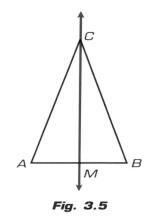

Fig. 3.5

Given this information, we may deduce several other properties of an isosceles triangle:

1. Since $r(A) = B$, $\overleftrightarrow{CM}$ is the perpendicular bisector of $\overline{AB}$.
2. Since $r(A) = B$, M is the midpoint of $\overline{AB}$ and $\overline{CM}$ is a median.
3. Since $r(C) = C$, $r(A) = B$, and $r(M) = M$, $r(\angle CAM) = \angle CBM$.
4. Since $r(\angle CAM) = \angle CBM$, $m\angle CAM = m\angle CBM$.

The triangle with rotational symmetry for magnitudes less than 360° is the equilateral triangle. If we rotate it 120°, 240°, or 360° about its centroid, the equilateral triangle will be its own image. This symmetry allows us to conclude that all the angles are congruent and all the sides are of equal length. Thus it is a regular polygon.

The symmetry of the isosceles triangle suggests a particularly useful placement of the triangle on the coordinate plane. It is shown in figure 3.6.

Notice also that reflection over the y-axis of A is B and the reverse. The coordinates of the points are (a, 0) and (–a, 0). That is, under reflection over the y-axis, the x-coordinates of the images are opposites and the y-coordinates are identical. In symbols, under a reflection in the y-axis,

$$(x, y) \longrightarrow (-x, y).$$

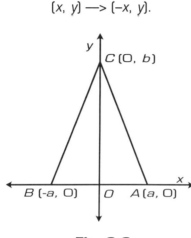

Fig. 3.6

Not only is the triangle a rigid shape, but a triangular region used repeatedly will cover the plane without overlapping triangles and without leaving holes. Thus a bathroom floor could be covered with triangular tiles. A region that will cover the plane as described above is said to *tile* or *tessellate* the plane. Complex tiling patterns arise in decorative arts such as carpets, fabrics, walls, floors, pottery, and baskets.

The tiling in figure 3.7a is called *edge-to-edge* because entire edges fit with entire edges. If we included non-edge-to-edge tilings as in figure 3.7b, a given figure that tesselates would have an infinite number of different patterns. In what follows, only edge-to-edge tilings are considered.

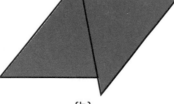

(a)	(b)
Edge-to-edge tiling	Non-edge-to-edge tiling

Fig. 3.7

Students should investigate the tiling capabilities of triangles. The scalene triangle should be used so that corresponding sides and angles can be distinguished. One tessellation pattern is possible when no reflection

Try this: If A has coordinates (x, y), what are the coordinates of A′ under a reflection over—

1. *the x-axis;*
2. *the line y = x;*
3. *the line y = –x;*
4. *the line y = 4;*
5. *the line x = –2?*

Try this: Make fifteen copies of one scalene triangle and, before cutting them out, color one side red. Use the triangles with the red side up to tessellate the plane. Determine the number of patterns possible when all the red sides are up. Can you find another pattern of the triangles that will tessellate the plane, if reflections of some triangles are permitted?

image of a triangle may be used. The pattern may be extended by translating any one of the three parallelograms (1, 2; 2, 3; or 3, 4) repeatedly (fig. 3.8a). Relaxation of the "no reflection" restriction leads to another pattern. In the other pattern as shown in figure 3.8b, the four numbered triangles form a unit that may be translated repeatedly to cover the entire plane.

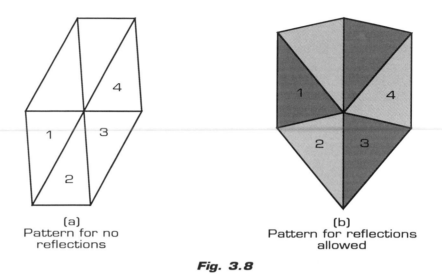

(a)
Pattern for no
reflections

(b)
Pattern for reflections
allowed

Fig. 3.8

Patterns that may arise when the tessellations begun in figures 3.8a and 3.8b are extended are shown in figure 3.9. An examination of the angles at the common vertex shows that each of the angles of the triangle must be represented twice. Since there are 360 degrees at a point, the angles of a triangle must sum to 180 degrees. Notice also that the "no reflections" pattern has no bilateral symmetry, whereas the "reflections allowed" pattern does. The "no reflections" pattern does have rotational symmetry—180 degrees about a common vertex.

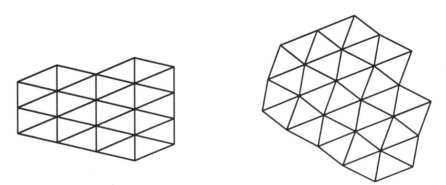

Fig. 3.9

The next several pages are blackline masters of activity sheets that can be used with geometry classes. They are designed to illustrate some of the ideas discussed in this chapter on triangles. They are also designed as investigative activities that may require more that one class period to complete. The first two illustrate the type of activity that can be based on use of the Geometric Supposer: Triangles, but other drawing and measuring utilities will work as well. The next two activities deal with coordinates and with generalization. Geometric Connectors: Transformations could be used in the third and Geometric Connectors: Coordinates could be used in the fourth. The fifth activity is a Geometry One investigation. The final activity features the chaos game and a Logo-based simulation.

ACTIVITY 1
GEOMETRIC SUPPOSER (TRIANGLES) INVESTIGATION:
MIDSEGMENT OF A TRIANGLE

1. **Supposer Construction:** Draw a triangle. *Subdivide* two sides of the triangle into two congruent segments. *Label* the midpoints and *draw* the segment connecting them.

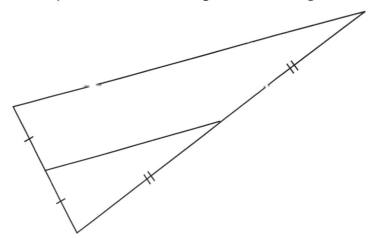

   Collect Data: *Measure* the length of the sides and the midsegment of the triangle. *Measure* the angles of the triangle and the angles formed by the midsegment and the sides of the triangle. Make a chart to record your data. Include columns for the type of triangle, the measures of the segments, and the measures of the angles. Be sure to leave room in the chart for additional entries. Neatly and carefully record your results.

2. Use the *repeat* command to do the Supposer Construction on a variety of triangles including right, obtuse, acute, isosceles, equilateral, and scalene. Collect the data noted above for each triangle. Complete the chart with the information obtained from the construction on these other triangles.

3. a. Looking at the lengths, do you see any patterns?

   b. Looking at the angle measures, do you see any patterns?

   c. Make your conjectures. Write them below.

   d. Justify your conjectures.

Note: *Italics* indicate a specific command on the Geometric Supposer.

ACTIVITY 2
GEOMETRIC SUPPOSER (TRIANGLES) INVESTIGATION:
LINE PARALLEL TO A SIDE OF A TRIANGLE

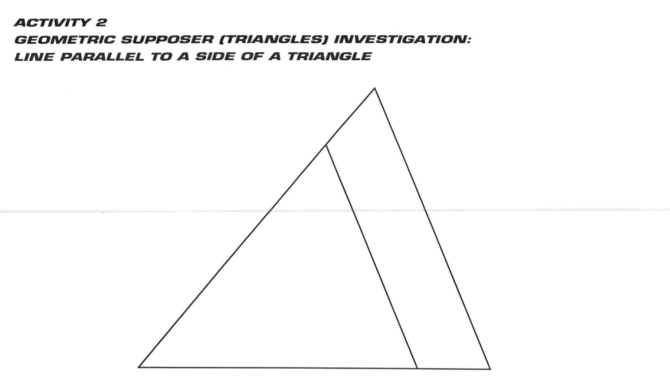

1. **Supposer Construction**: Draw a triangle. *Label* a random point on the triangle. *Draw* a line segment from the random point parallel to one side of the triangle and intersecting the other side, as in the diagram above. *Label* the intersection point.

   Collect Data: *Measure* the sides of the triangle and also *measure* the segments of the two sides determined by the parallel segment. Create a chart in which to record the data you collect.

2. *Repeat* this construction on a variety of triangles. Collect data on each triangle. Complete the chart that you began in #1.

3. Examine your data.

   a. Do you see any relationships among the lengths of the segments?

   b. Investigate sums, differences, ratios, and products. Write your conjectures.

   c. Try the Supposer Construction again to verify or refute your conjectures.

   d. Justify your conjectures.

Note: *Italics* indicate a specific command on the Geometric Supposer.

ACTIVITY 3
REFLECTIONS OVER THE LINE y = x ON THE COORDINATE PLANE

Points reflected over the line y = x are related to their preimages in a special way. Using coordinates makes the relationship especially clear.

1. On the following diagrams—

    a. Label the coordinates of the vertices for each figure.

    b. Sketch the reflection of each figure over the line y = x.

    c. Label the coordinates of the vertices of the image figures.

    d. What is the relationship between a point and its image over y = x?

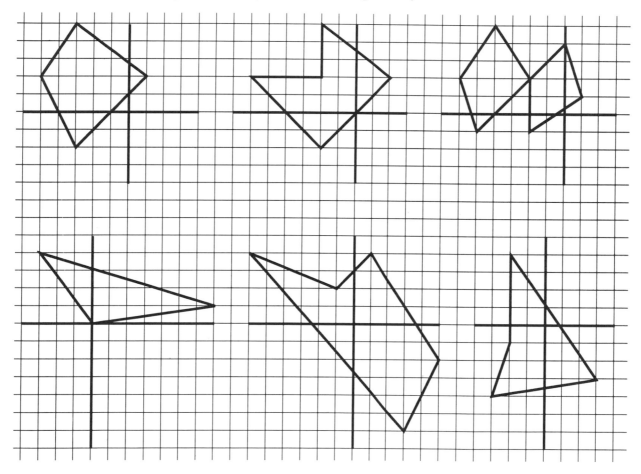

2. On graph paper, set up a coordinate system for each figure and graph the figure by plotting coordinates and connecting adjacent vertices. Sketch the reflection of each shape over the line y = x.

    a. (3, 2), (–1, –4), (7, 2), (–2, 3)

    b. (1, 7), (4, 5), (6, –1)

    c. (–2, –4), (–1, 5), (3, 3)

    d. (6, 4), (–2, 5), (–2, –2), (3, 5), (4, 2.5)

3. What is the relationship between the coordinates of the original figure and its reflection image? State your conjecture in if-then form. Write an argument that you could use with a friend to convince him or her that your conjecture is correct.

ACTIVITY 4
GENERAL PYTHAGORAS

1. A right triangle has a 90° angle in it. For convenience we will call the right angle *C* and the other two vertices *A* and *B* in right triangle *ABC*. On right triangle *ABC* below, construct squares on each leg and on the hypotenuse. Calculate the areas of each square. What relation exists among the areas?

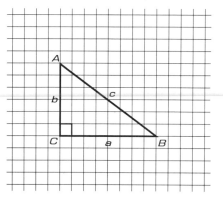

2. In #1 you confirmed the Pythagorean theorem that the sum of the areas of the squares on the legs is equal to the area of the square on the hypotenuse. On a sheet of triangular dot paper, construct right triangle *ABC*. Construct equilateral triangles on the legs and the hypotenuse. Compare the sum of the areas of the triangles on the legs to the area of the triangle on the hypotenuse. What do you find? Compare your findings with those of other students. State a conjecture in if-then form and justify it.

3. Draw right triangle *ABC* on a sheet of square dot paper. On each leg and the hypotenuse, construct an isosceles right triangle with congruent sides equal to the leg or the hypotenuse of right triangle *ABC*. Compare the sum of the areas of the triangles on the legs to the area of the triangle on the hypotenuse. State a conjecture in if-then form and justify it.

4. Draw right triangle *ABC*. On each leg and the hypotenuse, construct a semicircle. Compare the sum of the areas of the semicircles on the legs to the area of the semicircle on the hypotenuse. State a conjecture in if-then form and justify it.

5. Draw right triangle *ABC*. Construct a triangle of your choice on one leg. Construct similar triangles on the other leg and on the hypotenuse. Compare the sum of the areas of the similar triangles on the legs to the similar triangle on the hypotenuse. State a conjecture in if-then form and justify it.

6. On the basis of exercises 1–5 above, state a conjecture about figures constructed on the legs and hypotenuse of a right triangle. Choose a figure for which the conjecture has not yet been applied to see if your conjecture is supported. Explain why you think your conjecture is true.

ACTIVITY 5
GEOMETRY ONE INVESTIGATION: MIDPOINTS ON A TRIANGLE

1. Given an arbitrary triangle, $\triangle ABC$. The midpoints of each side of the triangle define a new triangle, $\triangle DEF$. It is common in geometry class to prove that $\triangle EFD$ is similar to $\triangle ABC$. Prove this.

2. In fact, we also can prove that triangles *ADF*, *DBE*, *FEC*, and *EFD* are all congruent. Complete the informal justification of this statement.

3. Since #2 is true, we should be able to generate $\triangle ABC$ when given $\triangle DEF$ by using the transformations of rotation, reflection, and translation of $\triangle DEF$.

   a. Use Geometry One to construct an arbitrary triangle.

EXPLORE: When finished press SKIP

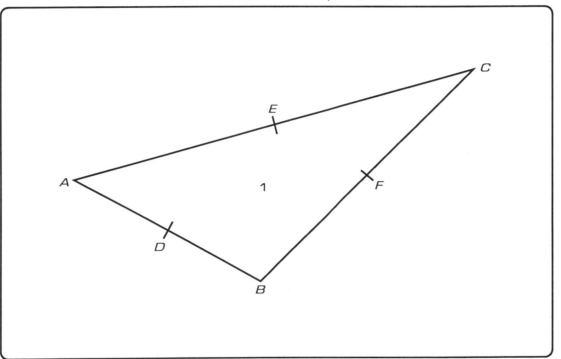

Translate Rotate rEflect Scale sHear
pUll Point N-gon Circle Define Keep

   b. Move to the transformation mode and define $\triangle ABC$ as triangle 1.

   c. Reflect, rotate, and translate triangle 1 to create a new triangle to which it is similar and for which the points *A*, *B*, and *C* are the midpoints of the new triangle.

   d. Sketch your result in the space below. Describe in a list of transformations the way that you generated the new triangle.

   e. Repeat the exercise to seek a different way to generate the triangle. What transformations did you use?

ACTIVITY 6
THE CHAOS GAME

Midpoints play an important role in developing a special fractal called the Sierpinski triangle. The triangle can be created by playing what has become known as the chaos game. Fix three points, A, B, and C as vertices of a triangle. The position of the points relative to each other is not particularly important, but if they form an equilateral triangle, the result of the chaos game is particularly pleasing. Pick any point X_0 as a starting point. You will use X_0 to find the next point X_1, and this point to find X_2, and so on. Now let's play the chaos game.

Rules of the chaos game:

1. Roll a die (or have a computer simulate a roll).

    a. If the die shows 1 or 2, plot the point that is the midpoint of segment $X_0 A$.

    b. If the die shows 3 or 4, plot the point that is the midpoint of segment $X_0 B$.

    c. If the die shows 5 or 6, plot the point that is the midpoint of segment $X_0 C$.

2. Call the midpoint just found X_1.

3. Repeat the process described in rules 1 and 2 using the new point X_1 as the starting point.

As you can see, the initial point X_0 creates a new point X_1, which, in turn, creates a new point X_2, which creates a new point X_3, which creates a new point,.... The question is, What is the graph of all the points X_n as n gets larger and larger?

1. Draw an equilateral triangle, $\triangle ABC$, and choose a point X_0 in the plane. Use a die and a ruler to locate at least 100 points in the sequence of Xs.

2. On a computer system supporting Logo, enter the following program and run it. Sketch the result of the run on your paper. What does the graph look like?

The following Logo program simulates 2000 plays of the chaos game.

```
(TERRAPIN LOGO 3.0)
TO CHAOS                TO SETUP                TO START
CS HT                   DOT -100 (-50)          MAKE "X (RANDOM 200)-100
SETUP                   DOT 100 (-50)           MAKE "Y (RANDOM 200)-100
START                   DOT 0 110               END
MAKE "COUNTER 0         END
PLAY :COUNTER
END


TO PLAY :COUNTER        TO MIDPOINT.A           TO DOT :X :Y
MAKE "Z RANDOM 3        MAKE "X (:X - 100)/2     PU SETXY :X (:Y)
IF :Z = 0 MIDPOINT.A    MAKE "Y (:Y - 50)/2      PD FD 1 PU
IF :Z = 1 MIDPOINT.B    DOT :X :Y                END
IF :Z = 2 MIDPOINT.C    END
IF :COUNTER = 2000 STOP
PLAY :COUNTER + 1


TO MIDPOINT.B           TO MIDPOINT.C
MAKE "X (:X + 100)/2    MAKE "X (:X)/2
MAKE "Y (:Y - 50)/2     MAKE "Y (:Y + 110)/2
DOT :X :Y               DOT :X :Y
END                     END
```

CHAPTER 4
QUADRILATERALS FROM MULTIPLE
PERSPECTIVES

The definitions of the six special quadrilaterals determine a hierarchy with a completely general quadrilateral at the top and the most specialized quadrilateral, the square, at the bottom (fig. 4.1). In this organization, the trapezoid is defined as a quadrilateral with *at least one* pair of opposite sides parallel. Thus a parallelogram is also a trapezoid; this allows the trapezoid to be included in the hierarchy, rather than sitting out by itself when the "*exactly one* pair of sides parallel" definition is chosen.

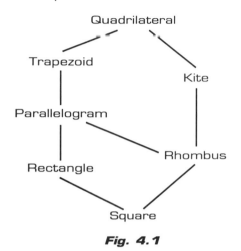

Fig. 4.1

The strength of choosing definitions that allow this hierarchy to be confirmed and justified is that when a theorem is proved for one shape, the same theorem holds for each shape below and connected to the given shape. For example, a kite has perpendicular diagonals. Thus a rhombus and a square also have perpendicular diagonals. A parallelogram has diagonals that bisect each other. It follows that a rhombus, a rectangle, and a square have the same property. Reasoning is efficient!

Students must be able to sketch representative examples of each quadrilateral when it is named. Sketches of kites should not appear to be rhombi, sketches of rectangles should not appear to be squares, and so on. The sketches used by students when they are given a verbal problem may affect the students' reasoning. Thus the sketches should not hint at properties of quadrilaterals lower in the hierarchy.

Once the quadrilaterals are introduced, they can be represented on the coordinate plane. As with triangles, the usual convenient placement is with one vertex at the origin and one side along the positive x- or y-axis. This certainly is true for all trapezoids (trapezoids, parallelograms, rectangles, and squares). For example, when a parallelogram is placed as in figure 4.2, the indicated coordinates are appropriate. Notice that, since $\overline{QA} \parallel \overline{ST}$, both S and T have the same y-coordinate. Further, since $OA = ST$, we may rename the coordinates of T as follows: S is b units from the y-axis, so T is a units farther from the y-axis. That is, $d = b + a$ and T is $(b + a, c)$. This sort of assignment for the coordinates is necessary when we begin to derive or justify properties using coordinate methods. Of course before that is done, examples using numerical coordinates should be tried. A good example that requires students to visualize the vertices in a variety of positions is to ask them to find the fourth vertex for each of three parallelograms with vertices of

Teaching Matters: A simple way to check whether sketches are general enough is to ask each student to sketch a variety of quadrilaterals and to measure the sides and angles. If the measures approximate those of quadrilaterals lower in the hierarchy, practice in sketching is needed.

Try this: With a vertex at (0, 0) and a side on the x-axis, find the most informative coordinatization of the remaining three vertices of a rectangle, a square, a rhombus, and an isosceles trapezoid. Do the same with the intersection of the diagonals at (0, 0).

Teaching Matters: Students should begin to reason using coordinate representations of geometric figures. For example, if the coordinates of a quadrilateral are (0, 0), (a, 0), (0, b), and (c, b), what can they conclude about this figure? They should justify their conclusions. (Two sides are parallel, since there are two sets of coordinates with the same y-coordinate.) Further reasoning can be encouraged by asking students to identify relationships among the coordinates that ensure a variety of quadrilaterals.

Assessment Matters: Students may have difficulty with coordinatization of special quadrilaterals because they do not understand the properties of coordinates, for example, that (b, c) and (d, c) are different points on the same horizontal line. Do not assume without further evidence that errors on homework or tests necessarily mean a lack of understanding of the properties of quadrilaterals.

(0, 0), (2, 1), and (1, 3). Several solution procedures are appropriate, but one that clarifies the question is to think of each pair of the given points as defining a diagonal. Then we will see that the required points are (−1, 2), (1, −2), and (3, 4).

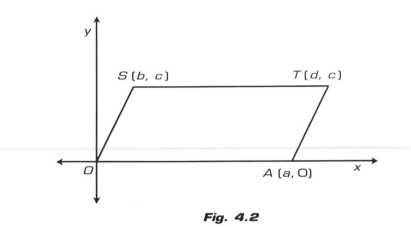

Fig. 4.2

Since a kite and its more specialized relatives have perpendicular diagonals, these figures are conveniently placed with the diagonals intersecting at the origin. Similarly, the parallelograms have bisecting diagonals, so they may be placed with that point at the origin and with the vertices coordinatized accordingly. The various placements of quadrilaterals on the coordinate plane should make clear the arbitrary nature of the placement and highlight how useful the properties of each figure are in assigning coordinates to vertices.

In a manner similar to that suggested for triangles, drawing and measuring utilities are excellent tools to use for studying the properties of quadrilaterals. The "redraw" capability of some of this software permits a construction done on one quadrilateral to be easily done again on a different quadrilateral. The user simply presses the space bar repeatedly. For example, suppose we had the computer draw a rectangle with the midpoints of each side constructed using the Geometric Supposer: Quadrilaterals or the Geometric Connectors: Coordinates. We then connected these points and noted that the resulting quadrilateral was a square. This software allows the construction done on the rectangle to be repeated on a rhombus or on any other quadrilateral. The conjecture made for the rectangle can thus be tested for other quadrilaterals and modified as additional information is obtained. The Geometer's Sketchpad allows the quadrilateral to be modified "on screen," and all measurements of angles, segments, areas, and so on, are immediately updated. Conjectures can be checked rapidly!

Recall that in a plane, if a line is perpendicular to one of two parallel lines, it is perpendicular to the other. Once this fact is established, we may demonstrate that various quadrilaterals have bilateral symmetry or are reflection symmetric. There are two main theorems here:

1. *An isosceles trapezoid is reflection symmetric with respect to the perpendicular bisector of its base.*

2. *A kite is reflection symmetric with respect to the diagonal that is the bisector of the second diagonal.*

Proof of theorem 1. In figure 4.3, $l \perp \overline{DC}$, $DM = MC$, and $m\angle DMN = m\angle CMN = 90°$. Thus $r(D) = C$ and $r(C) = D$. l is also perpendicular to segment AB, which implies r(line AB) is line AB. Since $m\angle D = m\angle C$, the reflection of ∠D is ∠C and the converse. Under a reflection, an intersection in the preimage is mapped onto the intersection of the images. Since A is the intersection of ray DA and line AB, which have images ray CB and line AB, r(A) = B and, conversely, r(B) = A. Thus r(ABCD) = BADC and the isosceles trapezoid is reflection symmetric.

*Try this: **Draw the quadrilateral A(–2, –1), B(1, 4), C(4, 3), and D(2, –4). Find the midpoints of each side and connect them in order. Investigate the resulting quadrilateral. Can you identify it? Use a drawing and measuring utility to construct a series of quadrilaterals with connected midpoints. Does the result for the specific example remain true? Generalize this result with vertices (x₁, y₁), (x₂, y₂), (x₃, y₃), and (x₄, y₄) and prove your generalization.***

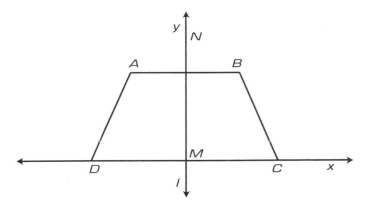

Fig. 4.3

This theorem, along with the quadrilateral hierarchy, again demonstrates the efficiency of a logical organization. Note that even though a parallelogram is *not* an isosceles trapezoid, the rectangle and the square are. Thus they also have bilateral symmetry—in fact, they have two symmetry lines, the perpendicular bisector of a side and the perpendicular bisector of an adjacent side.

The proof of the symmetry of the kite is left for a classroom exercise. If we begin with the perpendicular bisector of a diagonal, the argument is easy. Once we have this fact, we can conclude that rhombi and squares are reflection symmetric in each of their diagonals. And in the case of the square, it has four lines of symmetry.

Now recall that rotations are composites of reflections in intersecting lines and that the magnitude of the rotation is twice the angle between the reflecting lines. On one hand, a rectangle has 180° rotational symmetry because its lines of bilateral symmetry are perpendicular. A rhombus has this same property, but it is because its diagonals are perpendicular. The square, on the other hand, has rotational symmetry of 90°, 180°, and 270°, since its symmetry lines meet in angles of 45°, 90°, and 135°.

*Try this: **With software that will draw parallelograms, draw at least eight different parallelograms. In each case, make a table of the measures of the angles, the sides, and the diagonals. Look for possible relations among these measures as well as for possible relations among the triangles formed by the diagonals. State and justify as many conjectures as you find.***

*Try this: **Use a drawing and measuring utility to investigate the properties of parallelograms. Gather and organize data, and justify your conjectures.***

Quadrilaterals may also be used to tile (tessellate) the plane. The tilings we shall discuss are edge-to-edge tilings, that is, entire edges pair with entire edges only (fig. 4.4). This restriction limits the number of tilings that are possible with a given shape.

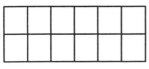

Edge-to-edge tiling

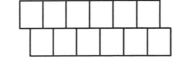

Non-edge-to-edge tiling

Fig. 4.4

The obvious tilings include those with squares and rectangles. Examples are floor and ceiling tiles. The question is whether trapezoids, parallelo-

Try this: Investigate the reflection and rotational symmetry of a parallelogram. Does it have reflection symmetry? Does it have rotational symmetry? Can you find a pair of reflecting lines that can be used to get the half-turn rotational symmetry?

grams, kites, and general quadrilaterals tessellate. These questions make good problems for students to investigate. Not only should they determine if a shape will tessellate the plane, but they should seek essentially different ways to accomplish the tiling.

Consider the case of a nonisosceles trapezoid (see fig. 4.5). Since the tiling must be edge to edge and no sides are congruent, we must find ways to move *ABCD* around the plane so that it will cover the plane. When we say "move around the plane," we mean to reflect, to rotate, or to translate the given shape. For the given trapezoid, translation will not work, since a segment and its translation image are both parallel and congruent (we have no congruent *and* parallel sides in *ABCD*).

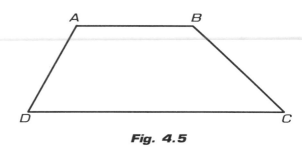

Fig. 4.5

We now ask, Can *ABCD* be rotated so that it will tile the plane? If rotation is to be used, it must result in an edge-to-edge pattern. The only rotation that will do that is a half-turn about the midpoint of a side. The four possible two-figure configurations that result from half-turns of *ABCD* are shown in figure 4.6.

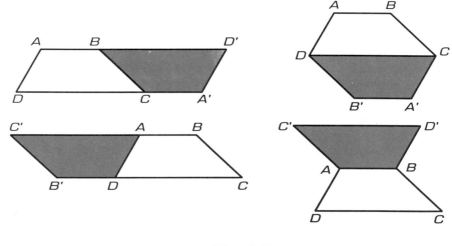

Fig. 4.6

Assessment Matters: To assess the learning outcomes of activities like those described in this section, have students work in small groups on a similar activity and observe the individuals in each group to see how much they contribute. Then have each student write an individual description of the activity, the conjectures that were made, and an argument for the validity of one or more conjectures. You might also give both a group and an individual grade, so that all students are motivated to contribute to the group effort.

Each of these two-trapezoid shapes may be moved by translations to completely tile the plane. Students should actually perform the motions to see the patterns. If the final patterns are compared for these four seemingly different beginning patterns, students will see that they are identical. In this pattern, if we began with paper colored on one side, we would have a pattern of one color only. This is because the entire pattern was made by rotations and translations only. No single reflections were used. However, if we reflect over one of the two parallel sides, and then half turn the two-trapezoid shape about the midpoint of one of the nonparallel sides, the four-trapezoid figure will tessellate the plane by translating it. See figures 4.7 and 4.8. The translating vectors for figure 4.7 are $\overrightarrow{AA'}$ and $\overrightarrow{B'''C}$. For figure 4.8, they are $\overrightarrow{C''B}$ and $\overrightarrow{D''A}$.

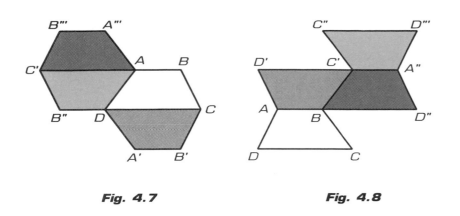

Fig. 4.7   **Fig. 4.8**

The remainder of the quadrilaterals will also tessellate the plane. Some, such as a parallelogram, need be moved only by translation. Others, such as a kite and a general quadrilateral, need half-turns to produce a larger "tile" that can be translated throughout the plane to cover it.

Several classroom-ready activities follow. They are designed to illustrate the richness of the possible investigations associated with quadrilaterals in geometry. As with the previous activities, these may take more than a single class period to complete. Also several of these activities may be the subjects of investigation by small groups of students. If this procedure is used, make sure that all students participate and are assessed on their knowledge and work.

Try this: **Given ABCD as in figures 4.7 and 4.8, have the students find a tessellation for which turning over a shape is permitted. That is, have them find the way to tessellate that is described at the bottom of page 26. Start with shapes colored on one side so that the flips can be identified. Find the smallest number of trapezoids needed to permit complete covering by simple translations.**

Try this: **Begin with a general quadrilateral with no sides parallel and no sides congruent. Make at least twelve copies of the shape on paper colored on one side. Search for a shape, made of the quadrilaterals, that will tile the plane by translation alone.**

Try this: **Run the following Logo procedure for regular polygons with three, four, and six sides. What does the program do?**

```
TO POLYGON :NUMSIDES :LENGTH
LT 90
MAKE "TURN 360/ :NUMSIDES
REPEAT :NUMSIDES
   [FD :LENGTH RT :TURN]
END
```

ACTIVITY 7
GEOMETRIC SUPPOSER (QUADRILATERALS) INVESTIGATION: MIDPOINTS OF SIDES OF A QUADRILATERAL

After reading and thinking about the following instructions, make a Supposer Construction that will enable you to use the *repeat* command. Quadrilaterals that will help you to investigate each situation are described below.

Create a quadrilateral of your choice. Subdivide each side into two segments and label the midpoint of each side. Connect midpoints on adjacent sides so that another quadrilateral is formed. Measure the lengths of the sides and the sizes of the angles to help you determine the nature of the quadrilateral formed.

Continue the investigation of quadrilaterals with several examples of each type of quadrilateral. Neatly and carefully record all your data for later analysis.

What figure is formed if the original quadrilateral is a parallelogram? A trapezoid? An isosceles trapezoid? A kite? A square? A rhombus? A rectangle?

What figure is formed when the consecutive midpoints of any kind of quadrilateral are connected?

Can the quadrilateral formed by joining the midpoints be a rhombus? A rectangle? A square? Write your conjectures. Test your conjectures with the Supposer. Justify your conjectures.

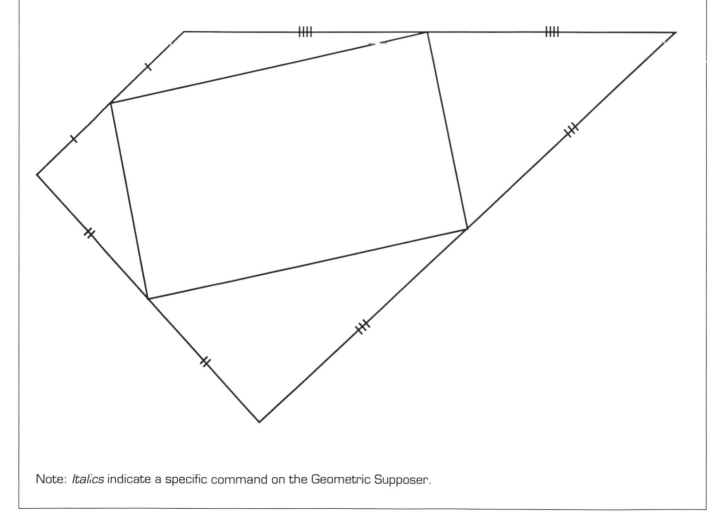

Note: *Italics* indicate a specific command on the Geometric Supposer.

ACTIVITY 8
AREAS OF PARALLELOGRAMS

Instructions: You will need a geoboard, several rubber bands, materials to record your discoveries, and a partner. You will need to use the formulas for the areas of a square, a rectangle, and a triangle.

Recall that on a geoboard, $\vert$ has length 1, whereas $\diagdown$ has length $\sqrt{2}$.

1. Find the area of each parallelogram by using your geoboard. Subdivide each figure into triangles, squares, and rectangles to help you. For each parallelogram, record the following information in a table: length of base height, and area.

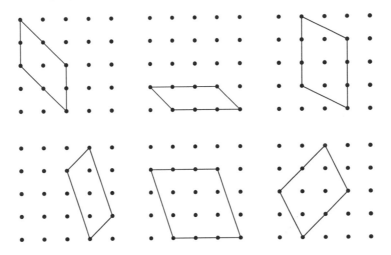

2. Examine your table. Can you see a relationship between the length of the base, the height, and the area of the parallelogram? State your conjecture and discuss it with the other members of the class. Justify the formula you discovered.

3. You can use a similar approach to develop a formula for the area of a trapezoid. Make a table to keep track of the lengths of both bases, the height of the trapezoid, and the area. Use the following trapezoids as a start.

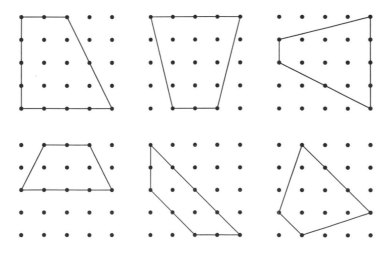

4. Write your formula for the area of a trapezoid. Write a justification for your formula.

ACTIVITY 9
EQUILIC QUADRILATERALS

1. You know many quadrilaterals such as the square, the rhombus, the trapezoid, and the parallelogram, but have you ever heard of an *equilic quadrilateral?* An equilic quadrilateral is a quadrilateral with a pair of congruent opposite sides that if extended, meet to form a 60° angle. The other two sides are called *bases.* Using any drawing or construction tools you wish, draw an accurate equilic quadrilateral on another sheet of paper.

2. Under your equilic quadrilateral write the sequence of steps that you used to draw it. Compare your procedure with that of a neighbor. How do they differ? Do they describe the same procedure? Work together to clarify your descriptions so that a classmate could use them to draw an equilic quadrilateral. Write them down.

3. In the rest of the investigations on this sheet we label any equilic quadrilateral *ABCD* where $\overline{AB}$ is the shorter base and $\overline{CD}$ is the longer base. $\overline{AC}$ and $\overline{BD}$ are diagonals. Answer the following questions for equilic quadrilateral *ABCD:*

   a. Can $\overline{AB}$ be parallel to $\overline{CD}$? If so, draw a model; if not, explain.
   b. Can $\overline{AB}$ be congruent to $\overline{CD}$? If so, draw a model; if not, explain.
   c. Can $\overline{AB}$ or $\overline{CD}$ be congruent to $\overline{BC}$ and $\overline{DA}$? If so, draw a model; if not, explain.
   d. Can $\overline{AC} = \overline{BD}$? If so, draw a model; if not, explain.
   e. Find the sum of the measures of angles *A* and *B;* of angles *C* and *D*.
   f. Can angle *C* or *D* have measure 90°? If so, draw a model; if not, explain.

4. Below is equilic quadrilateral *ABCD* with diagonals $\overline{AC}$ and $\overline{BD}$. Construct the midpoints of $\overline{AB}$, $\overline{CD}$, $\overline{AC}$, and $\overline{BD}$, calling them *J, K, L,* and *M* respectively. What, if anything, can you say about the relationship of the four midpoints or about the relationship of any set of three midpoints? State your conjectures in if-then form and justify them.

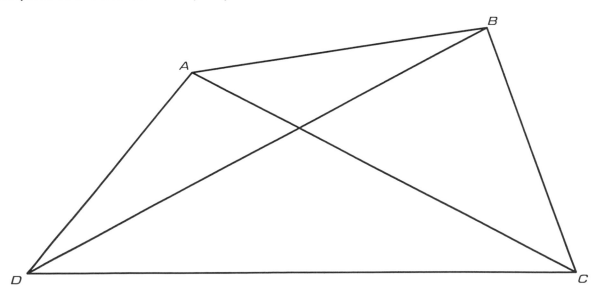

5. Draw equilic quadrilateral *ABCD*. Draw an equilateral triangle with base $\overline{AB}$ and third vertex on the side of $\overleftrightarrow{AB}$ opposite to that in which *C* and *D* lie. Call this vertex *P*. How are the points *P, C,* and *D* related? State your conjecture in if-then form and justify it.

ACTIVITY 10
QUADRILATERAL INVESTIGATION

1. If you were to draw the bisectors of the angles of a square, they would meet in a single point because they are the diagonals. However, draw a rectangle and construct the bisectors of the four interior angles using a Mira, compass, or paper folding, as described below. These four lines intersect in four points, *W, X, Y,* and *Z*, as shown below. List as many properties of these four points as you can. Compare your list with that of a partner. With a partner, justify each property in your combined lists.

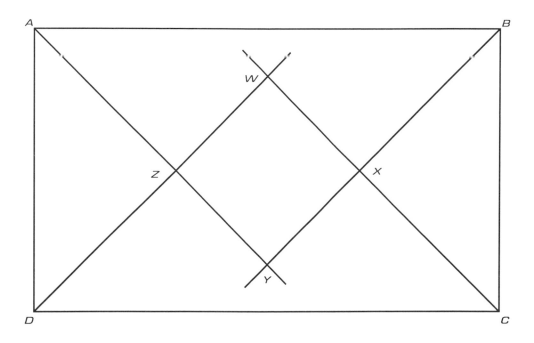

In Exercises 1 and 2, the instructions call for you to construct bisectors of the angles of a quadrilateral. You may do this by using a compass, a Mira, or paper folding. To use a Mira, construct the required quadrilateral, place the Mira so that one side of a vertex angle reflects onto the other side, and draw the reflecting line. In the figure above, the lines *AY, DW, BX,* and *CW* are the reflecting lines. To use paper folding, cut out the required quadrilateral. Make a fold through a vertex, creasing the paper so that the two sides of the angle fall on each other. The lines identified as reflecting lines are the creases.

2. Carry out the procedure described above, using any of the three construction methods you wish, for each of the quadrilaterals identified below. Make a list of the properties of the four points of intersection for each quadrilateral:

   a. A rhombus

   b. A parallelogram

   c. A kite

   d. An isosceles trapezoid

   e. A trapezoid

   f. A general quadrilateral, that is, a quadrilateral with no special characteristics

3. Examine the lists of properties you generated in parts a–f of Exercise 2. Are there any properties of the four points constructed that are common to *each* of the quadrilaterals? If so, write your conjecture in if-then form and write a justification for it.

CHAPTER 5
POLYGONS FROM MULTIPLE PERSPECTIVES

In addition to triangles and quadrilaterals, the regular polygons are a significant topic in geometry. These shapes have many symmetries. The investigation of these symmetries is an important topic for school geometry. Students should note that for even-sided regular polygons, the bilateral (reflection) symmetry lines either contain two vertices or are perpendicular to two parallel sides. (See fig. 5.1.) In the case of odd-sided regular polygons, there is only one kind of symmetry line, namely, one that is perpendicular to a side *and* contains a vertex.

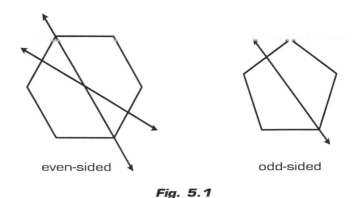

even-sided   odd-sided

Fig. 5.1

Teaching Matters: Paper folding may be used to determine the sum of the measures of the interior angles of a triangle. Simply fold to find the midpoints of two sides and then fold along the line determined by these points and along the lines perpendicular to the third side from these midpoints. This information may be used to develop a formula for the sum of the interior angles of any convex polygon of n sides.

Under rotation, regular polygons exhibit rotational symmetry. The order of the rotational symmetry group is the same as the number of sides of the polygon. Again, the relation of the magnitude of the rotation and the angle between reflecting lines needs to be noted.

In three dimensions, polyhedra are used to model the crystal structures of elements and compounds. Until recently it was thought that only rotations of 180°, 120°, 90°, and 60° could produce identical crystals. In an attempt to develop a very strong alloy, experimenters worked to produce an alloy of manganese and aluminum. This was very difficult to do because the process required a rapid decrease in temperature. In 1982, chemist Daniel Shechtman began a series of tests to determine the crystalline structure of the small quantities of alloy produced. To his amazement and to the disbelief of the scientific community, he found fivefold rotational symmetry. Yet further investigation confirmed Shechtman's analyses. A crystal of another aluminum alloy is shown in figure 5.2. You can see the fivefold rotational symmetry in the center of the figure (Steen 1988, p. 265).

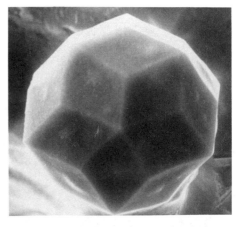

Try this: Consider the first ten regular polygons. Make several replicas of each. By experimentation, determine which can or cannot be used to tessellate the plane. In groups, develop a justification that only the three-, four-, and six-sided regular polygons can be used to tessellate.

An interesting question concerning regular polygons is, Which of them will tessellate the plane in an edge-to-edge manner? It is easy to show that only those whose interior angle measure divides 360 evenly will tessellate.

Fig. 5.2

There is a video entitled "On Size and Shape: Scale and Form," tape number 17 in the Annenberg/CPB series *For All Practical Purposes* (available from 901 E Street, NW, Washington, DC 20004–2037), which gives an excellent discussion on tilings and the conditions for regular polygons to tile the plane. Also included on this tape is a discussion of Roger Penrose's tiles and the fivefold symmetry of the manganese-aluminum alloy.

There is a great deal more that could be done with tessellations and symmetry to modernize school geometry, but we shall introduce only three other topics. Both symmetry and tessellation are found in the art of many peoples. Perhaps the most famous art is that of M. C. Escher, a Dutch artist whose work has been displayed worldwide. Figure 5.4 is an example of an Escher, with a graphic demonstration of how it could be developed from a square (fig. 5.5). A reference for this type of work is Ranucci and Teeters's *Creating Escher-Type Drawings* (1977, p. 49).

*Try this: **Construct multiple copies of Roger Penrose's kite and dart using the pattern in figure 5.3 (adapted from Steen 1988, p. 263). Use the kite and dart to tessellate the plane. Try to make a tessellation in which the "kite + dart" rhombus below is not a part. Does the pattern generated have any rotational or bilateral symmetry? If so, describe it.***

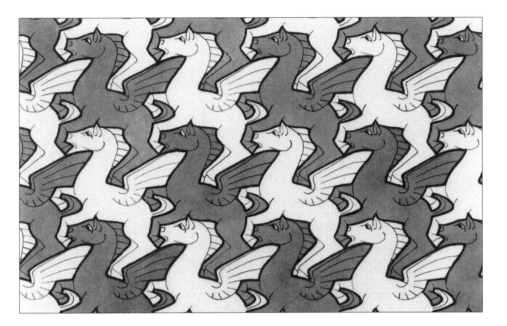

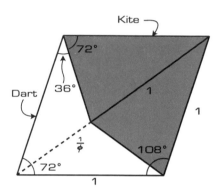

Fig. 5.3

Fig. 5.4. M. C. Escher periodic drawing

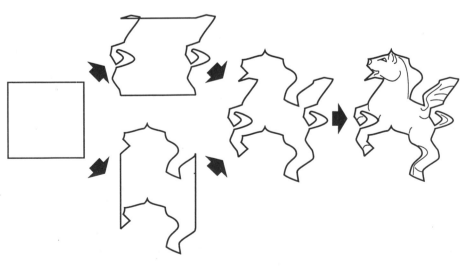

*Try this: **Have students use the technique outlined by Teeters (1974) to make tessellation patterns.***

Fig. 5.5

Escher's work is unique, but it does not represent the bulk of the artwork that may be analyzed and explained using geometric transformations. The first is the *strip* or *frieze pattern*. It is represented by decoration around pottery and by the artwork in cornices on buildings.

Figure 5.6 is an example of a strip pattern. It is considered to be infinite, and a translation will map the pattern onto itself. It is said to have *translation symmetry*. Such symmetry is characteristic of all strip patterns.

...Ç Ç Ç Ç Ç Ç Ç Ç...

Fig. 5.6

A strip pattern with translational symmetry may also be symmetric under reflection, rotation, or glide reflection. These are illustrated below.

1. Reflection over a horizontal line

```
b b b b b b b b
<————————————————>
p p p p p p p p
```

2. Reflection over a vertical line

```
              ^
...q p  q p  q p | q p  q p  q p...
              v
```

3. Half-turn about a point on the midline of the strip

```
q q q q q q
<——————•——————>
  b b b b b b
```

4. Reflection over a horizontal line followed by a translation in the direction of the line. This is a glide reflection.

```
   q   q   q   q
<——————————————————>
 d   d   d   d
```

There are only seven such patterns possible. You have seen four of the seven. The remaining three are given below.

5. The only symmetry is translational

```
q   q   q   q
```

6. Reflection over a vertical line and a glide reflection along the midline

```
   bd    bd    bd    bd
<——————————————————————>
q    pq    pq    pq    p
```

7. Reflection over a vertical line and reflection in the horizontal midline

```
d b   d b   d b   d b
<——————————————————————>
q p   q p   q p   q p
```

***Teaching Matters:* Strip patterns are found in wallpaper, in artwork, in cornices of structures, and in other areas. Students can be asked to find such patterns, sketch them, and classify them. Another idea is to have each student construct a pattern following one of the patterns at the left and then to put them on the bulletin board. Such an activity serves to underscore the contribution of geometry to our cultural heritage.**

Mathematicians have developed a notation to denote each strip pattern (Crowe and Thompson 1987, p. 108). Each symbol is made up of two characters. The first character is "m" or "1" according to whether there is a reflection over a vertical line or not. The second character is "m" if there is a reflection over a horizontal line, "g" if there is a glide reflection, "2" if there is a half-turn, and "1" if none of these exists. In these symbols the patterns shown above are as follows:

1 is 1m;	2 is m1;	3 is 12;	4 is 1g;
5 is 11;	6 is mg;	7 is mm.	

These patterns arise in the artwork of native cultures. Figure 5.7 is an example of the seven strip patterns from the pottery of San Ildefonso Pueblo, New Mexico.

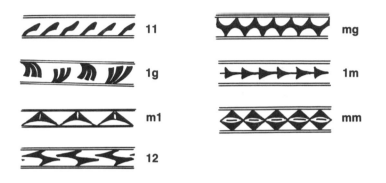

Reprinted from Crowe and Thompson (1987, p. 108).

Fig. 5.7

The seventeen possible two-color strip patterns are shown in figure 5.8.

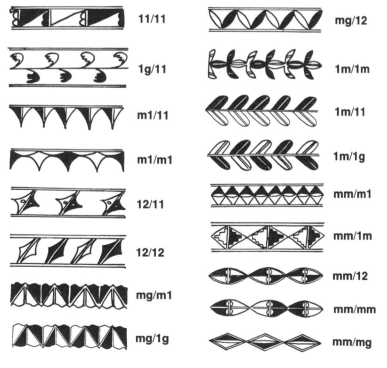

Reprinted from Crowe and Thompson (1987, p. 108).

Fig. 5.8

Try this: The strip patterns described at right are all one color. Suppose we can use two colors such that the pattern has translational symmetry, that is, the colored parts match when translated. Draw some strip patterns using two colors and classify the resulting patterns according to the rules given previously.

The final topic we will consider is an extension of the strip patterns. It is called a *wallpaper pattern*. In a wallpaper pattern, the pattern extends indefinitely both horizontally and vertically. Such patterns are found in the creative work of carpet makers, fabric weavers, and basket makers, to name a few. There are seventeen such patterns possible. The seventeen patterns are shown in figure 5.9, along with a scheme that can be used for categorizing a pattern. Many of these patterns are very complex and hard to determine. Further information on the mathematics of these patterns may be found in Steen (1988, p. 259) and Crowe (1986, p. 25).

Try this: Find a wallpaper that has a pattern in one of the forms in figure 5.9. Sketch it and classify it.

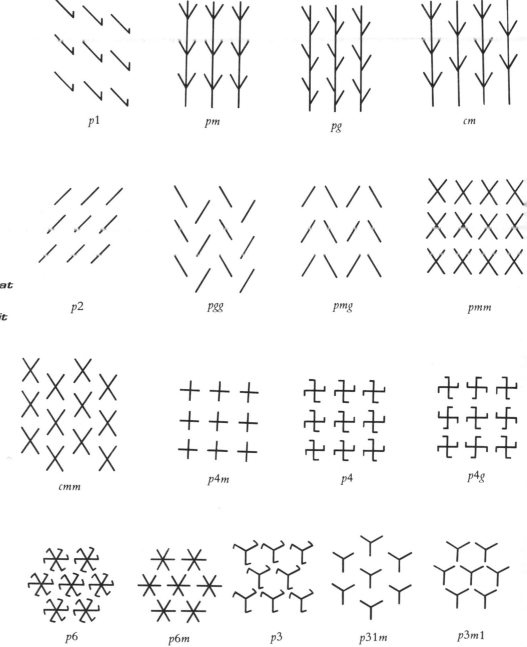

Fig. 5.9

The connections among wallpaper patterns, symmetry, and transformations are illustrated in figures 5.10 and 5.11.

Cotton printer's block, India (p1) [Stevens 1981, p. 178] reprinted by permission of MIT Press.

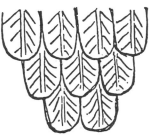

Carved feathers (cm), Benin, Nigeria [Crowe 1975, p. 258].

Illuminated manuscript (p6m), Persia [Christie 1969, p. 244].

Painted ceiling (p4), ancient Egypt [Christie 1969, p. 232].

Cotton printer's block (p4g), India [Christie 1969, p. 232].

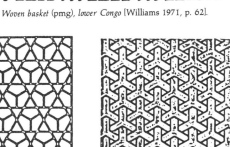

Woven basket (pmg), lower Congo [Williams 1971, p. 62].

(p3), ancient Egypt [Christie 1969].

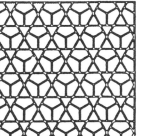

Cotton printer's block (pg), India [Christie 1969, p. 197].

Window lattice (p3m1), China [Dye 1974].

(p31m), China [Dye 1974].

(p6), Persia

Stamped cloth (pmm), Ghana [Williams 1971, p. 30].

Stamped cloth (p2), Ghana [Williams 1971, p. 30].

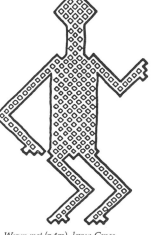

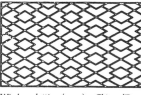

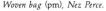

Carved cup (pgg), Bakuba [Zaslavsky 1973, p. 191].

Woven mat (p4m), lower Congo [Williams 1971, p. 95].

Window lattice (cmm), China [Dye 1974].

Woven bag (pm), Nez Perce.

Reprinted from Crowe 1986.

Fig. 5.10

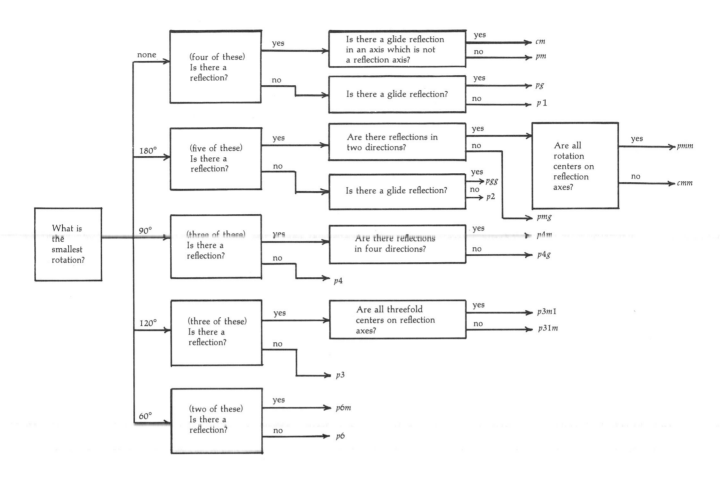

Reprinted from Crowe 1986.

Fig. 5.11. Flowchart for the seventeen two-dimensional patterns

The activity sheets that follow illustrate some types of activities that can be built into a geometry course by using the ideas of this chapter. These activities illustrate the applicability of geometric ideas in the real world. For example, archaeologists use the seven translation symmetries of the strip patterns as cultural markers to help them identify the culture from which pottery comes. This topic is discussed in the video "On Size and Shape: Overview," tape number 16 in the Annenberg/CPG series *For All Practical Purposes.*

ACTIVITY 11
THE CAIRO TESSELLATION

The construction described below yields a tile that is an equilateral, nonregular pentagon. This tile is the fundamental tile for a beautiful tessellation called the *Cairo tessellation*. Such patterns were often used to pave the streets in Cairo. Follow the steps to create the Cairo tile, and then use it to make a tessellation.

Step 1: On another sheet of paper, construct segment AB, with midpoint M.

Step 2: In the same half-plane, construct two 45° angles with vertex M. Call them $\angle AMX$ and $\angle BMY$.

Step 3: Locate C on ray MY such that C lies on the circle having center B and radius AB.

Step 4: Locate E on ray MX such that E lies on the circle having center A and radius AB.

Step 5: Locate D, the intersection of the circle having center C and radius AB and the circle having center E and radius AB.

Step 6: Pentagon $ABCDE$ is the Cairo tile. Recopy it and create the Cairo tessellation!

Questions for discussion:

1. Streets in Cairo were decorated with the tessellation you just made. Are there any tessellations in your city? Check the tiling patterns on the floor in your school, in the shopping mall, on the sides of buildings. For each tessellation that you find, sketch the fundamental tile and state the location where it was found.

2. Look carefully at your Cairo tessellation. It was created using a convex, equilateral, nonequiangular pentagon. However, there are certain *combinations* of these pentagons that form new fundamental tiles. Can you identify any such combination? What type of polygon is the new (larger) fundamental tile, and how many Cairo tiles form it? List as many possibilities as you can.

3. The five perpendicular bisectors of the sides of a Cairo tile meet at the *center* of the tile. Connect all centers of tiles that share a common side and observe the resulting design. You should notice a very interesting pattern—describe it!

Challenge: Using the points of the construction given above, what type of angles are $\angle DEA$ and $\angle DCB$? Give some evidence to support your conjecture!

ACTIVITY 12
ARCHIMEDEAN DUALS

When mathematicians refer to "Archimedean duals," they are *not* referring to an ancient sword fight or soccer match—instead, they are describing new patterns and tilings found in Archimedean tessellations. A tessellation is said to be *Archimedean* or *semiregular* if it contains two or more regular polygons and if each vertex in the tessellation is surrounded by the same type and number of polygons in the same circular order. For example, the tessellation shown in the diagram below is an Archimedean tessellation in which each vertex is surrounded by an equilateral triangle, two squares, and a regular hexagon.

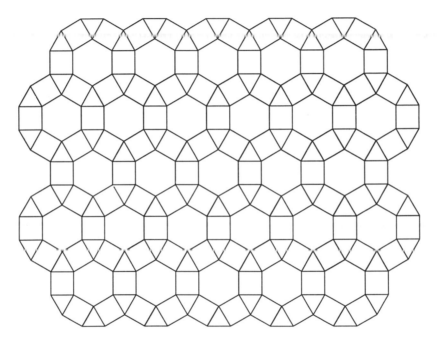

To create the *dual* of an Archimedean tessellation, connect the centers of all polygons that share a common side.

1. a. Use the diagram given above, and sketch in the dual of this tessellation.
   b. What pattern is formed when you examine just the dual?
   c. Is this new pattern a tessellation?

There are a total of eight Archimedean tessellations. The one shown in the diagram above is sometimes coded 3-4-6-4. This code tells us what *type* of regular polygons are being used (3 = triangle, etc.), how *many* of them there are, and in which circular *order* they appear.

2. a. Create two Archimedean tessellations involving three triangles and two squares. **Caution!** There are two! One is coded 3-3-3-4-4, and the other is coded 3-3-4-3-4.
   b. Sketch and label both of these and then draw in the *duals* for each pattern.
   c. What type of tiling is formed?
   d. Is it a tessellation?

Challenge: Can you determine the codes for the five remaining Archimedean tessellations? Explain your reasoning!

ACTIVITY 13
FRIEZE PATTERNS

The decorative horizontal bands that sometimes are found along the upper part of a wall are known as *frieze patterns.* Many of these artistic designs have interesting geometric properties, such as horizontal or vertical line symmetry, translational symmetry, and rotational symmetry. Mathematicians have classified the various types of frieze patterns into seven basic categories on the basis of the types of symmetries they possess. To classify a frieze pattern, we will assign a code, which has two symbols: the first will indicate whether the frieze pattern has vertical reflectional symmetry, and the second symbol will tell whether there is an additional property of symmetry. The following tree diagram shows this classification system, along with a simple example of each type of pattern:

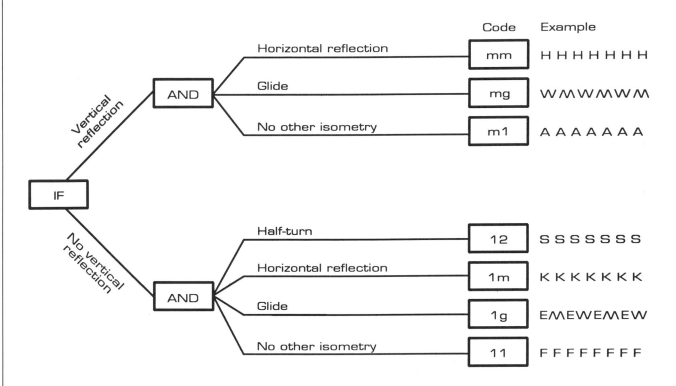

The examples shown above use letters of the alphabet, but more artistic designs are certainly possible!

1. For each of the seven categories of frieze patterns given in the tree diagram, design two patterns of your own: make the first fairly basic, using letters or numerals, and make the second pattern more complex, using some creative symbols of your own making.

2. Use a computer and a graphics software program to generate some frieze patterns. What special "tools" or options in the program are especially helpful in creating this type of pattern?

Challenge: What does the code m2 mean, and why was it not included in the tree diagram? Explain where frieze patterns having the properties of m2 would be found in the classification system we are using.

3. Using the codes for frieze patterns, classify each of the following designs:

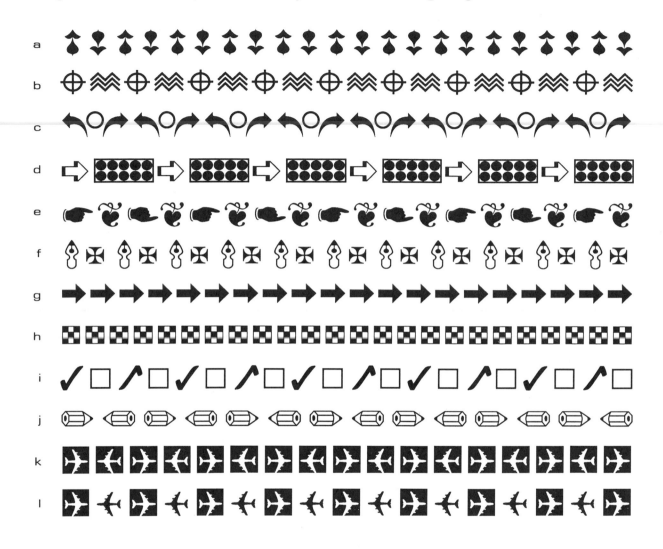

4. Examine buildings and rooms in your city for frieze patterns (older buildings often have this sort of decorative trim). Tell the location of the frieze pattern you find, sketch it, and then classify it using the codes given.

5. As a group activity, make a series of frieze patterns to decorate the wall in your mathematics class. You might even ask the principal to judge the different ones and make a contest out of it!

CHAPTER 6
SOLIDS: EXPANDED PERSPECTIVES

Solids such as prisms, pyramids, and spheres are often introduced late in the geometry curriculum. Their properties are briefly introduced, and surface area and volume formulas are derived. The *Curriculum and Evaluation Standards for School Mathematics* (NCTM 1989) suggests that this practice should be modified. Solids should be introduced with the shapes that describe their faces. For example, when triangles are studied, the tetrahedron should be introduced. Similarly when quadrilaterals are introduced, pyramids and prisms with quadrilateral bases should be introduced as well.

The major goals of the early introduction of polyhedra are to allow familiarity to develop and to improve the visualization and sketching skills of students. The former goal is often attained by emphasizing the latter goal. Thus a possible activity for tetrahedrons is to build models by using soda straws and string. The string is run through the straws and knotted to make the shape. Once the technique is established, students can be asked to construct special tetrahedra, such as one with six congruent edges, one with at least one face a right triangular region, one with a face perpendicular to the plane of the base, and so on. Similar sorts of activities can be designed for other solids as they are introduced.

Once straw models are available, another broadly useful activity is to ask students to draw a plane representation of the three-dimensional solid. This skill will be vital for work in the twenty-first century when computer graphic representations of solids will be common.

Drawing solids takes some instruction on how to represent edges that are not visible in opaque models. The standard procedure in mathematics is to dot those edges to provide the illusion of three dimensions. Thus in the sketch at left in figure 6.1, edge $\overline{BD}$ is at the rear of the tetrahedron, whereas at right, face *ABD* blocks the view of the remaining three edges.

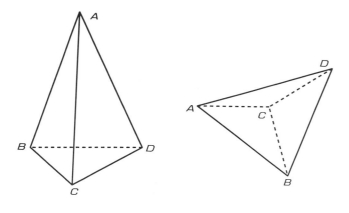

Fig. 6.1

Students should be asked to sketch solids from specific points of view, for example, when looking directly at the edge $\overline{AC}$ or the face *ABD* in figure 6.1. Also ask for sketches from the top, from the bottom, and from the inside of the solid.

Another good activity for increasing familiarity with solids is sketching the intersections of solids and planes. This should first be modeled with straw-and-string models. For example, in tetrahedron ABCD (fig. 6.1), a

Try this: Can a straw-and-string model of a tetrahedron be made without passing the string through a straw more than once?

Try this: Ask each student to determine the volume and surface area of the box of his or her favorite cereal. How does the volume change with the surface area? With the shape of the box? How does the volume of cereal differ from the volume of the box? Discuss the question of why the box is so large for the amount of cereal it contains.

Teaching Matters: Paper with points at the vertices of equilateral triangles is called isometric grid paper. It is useful for sketching three-dimensional shapes. Cubes are especially well represented on such paper.

Try this: Given a tetrahedron ABCD, sketch it while looking at—

a. A from 12 inches above it;
b. A from the center of the base BCD;
c. A from B.

Try this: **Given the "building" below, sketch its three elevations.**

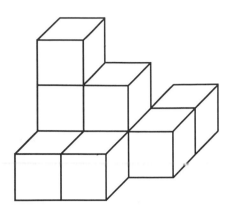

plane intersects sides $\overline{AB}$ and $\overline{AC}$ at their midpoints and intersects side $\overline{BD}$ at its midpoint. Sketch the intersection and identify the plane shape.

An extension of sketching solids from viewpoints is sketching the "elevations" of real objects. In architecture an elevation is a perpendicular projection of a building in a plane parallel to the face in question. Usually three elevations are drawn: a front view, a side view, and a top view. This architectural necessity is easily simulated by constructing "buildings" of cube blocks and asking students to sketch the three elevations. The reverse activity is also good for reasoning and visualization skills because most elevation sets permit several "buildings."

Another way to increase familiarity with solids is to draw *nets* of the solid. A net of a solid is a plane model of its faces that when cut out and folded up, will result in the solid. Thus a net of a cube is shown in figure 6.2. One question that could be raised is, How many nets are there for a given solid? Also, are there "apparent" nets that do not fold into the solid?

Try this: **Given six square regions, how many different edge-to-edge nets are possible? How many of these fold into a cube?**

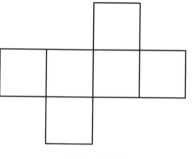

Fig. 6.2

The solids can also be investigated by using transformations. For reflection in three-dimensional space, the line is replaced by a plane, and we seek the plane of bilateral symmetry. For a rectangular parallelepiped (box), shown in figure 6.3, there are three symmetry planes. For rotational symmetry, the point that acts as the center of the rotation is replaced by a line that acts as an axis of symmetry. The box has three axes of rotation. They are the three lines of intersection of the pairs of planes. The angle of rotation is 180° in each case.

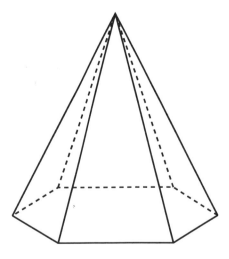

Try this: **Given a regular right hexagonal pyramid, determine the symmetry planes and the angle of rotational symmetry.**

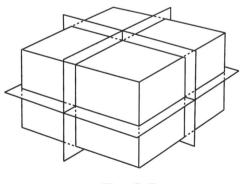

Fig. 6.3

Activities such as those that follow require students to envision transformed figures and the effects of motions like folding or turning on given figures. Such activities help develop a generalized ability called "spatial visualization," which is one of the abilities measured by standardized aptitude and cognitive ability tests. Do not be surprised if students vary *widely* in their spatial visualization ability.

ACTIVITY 14
VISUALIZING SOLIDS

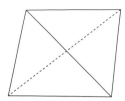

Can you create a tetrahedron using straws held together with string? Here's one way by Victoria Pohl (1987).

Choose six straws, all the same length. Thread a needle with about a meter of string. (Longer pieces tend to tangle; when the string becomes too short, tie on another piece.) Drop the threaded needle through three straws. Next tie the string to form a triangle, allowing no slack in the string (fig. 1). Drop the threaded needle through straws 4 and 5 (fig. 2). Drop the needle back through straw 2 and then through straw 6 (fig. 3). Pull the string tight. Drop the needle through straw 4 to return to the starting point (fig. 4). Tie the string securely and cut off the string ends.

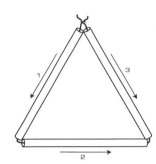

Fig. 1

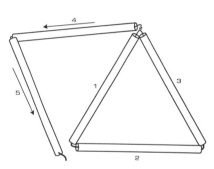

Fig. 2

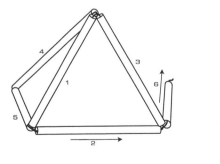

Fig. 3

Fig. 4

1. Is it possible to build a tetrahedron without taking the thread through any straw more than once? Explain your answer.

2. Using straws and string, build a cube. Build an octahedron. How many straws will you need for each? Can you take the thread through each straw only once for these figures?

Challenge: Can you build a tetrahedron inside a tetrahedron, a cube inside an octahedron, an octahedron inside a tetrahedron, or a tetrahedron inside a cube?

ACTIVITY 15
COLLAPSING CUBES

You can begin this investigation by trying to visualize how the sides of a cube can be collapsed into different two-dimensional patterns. First, we'll look at a simpler version of this problem. Suppose we look at a cube without one face—a box without a top. We can see that the box has only five sides and for simplicity, that these sides are all squares. If we unfold the box we might get

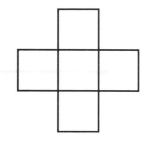 but we could not unfold the box to get the next pattern.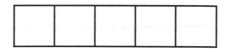

1. Find the other patterns that can be folded into a box without a top. Be careful not to count the same pattern twice. Said another way: rotations or reflections of a pattern are not counted a second time. For example,

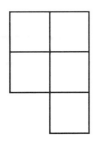 is the same as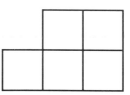

2. There are twelve different arrangements of five squares, but only the eight found in Exercise 1 can be folded up into a box without a top. Can you find all twelve? Are the areas of each of these the same? Explain! Are the perimeters of each of these the same? Explain!

3. Now see if you can collapse a regular tetrahedron in the same manner. You will have four equilateral triangles to work with. Again be careful not to count a rotation or a reflection of a combination you already have.

4. A challenging problem is to try to collapse a cube. This problem is very similar to the box-without-a-top problem. The only addition is one more square. Unfortunately, this one square increases the number of combinations to thirty-five. The number of combinations that are solutions is eleven. Find as many of them as you can. Also find as many of the six-square patterns as you can. Do these all have the same area? Perimeter?

CHAPTER 7
REASONING ABOUT SHAPES USING
COORDINATES AND TRANSFORMATIONS

A major goal in school geometry is the development of mathematical reasoning abilities. The *Curriculum and Evaluation Standards for School Mathematics* (NCTM 1989) reaffirms this goal but suggests that reasoning about shapes should use coordinate and transformation techniques as well as the traditional synthetic techniques. Theorems and related problems that involve parallelism, perpendicularity, equal distances, intersections, and midpoints are especially amenable to a coordinate approach. As these assertions arise, both synthetic and coordinate arguments should be given or assigned.

One approach to using coordinates would be to develop ideas from both a synthetic and a coordinate perspective simultaneously. In this development, coordinate representations could serve as a mathematical model of geometric ideas. The key aspects of this development follow (Hirsch et al. 1990).

1. A point is an ordered pair of real numbers (x, y).
2. A line is all pairs of ordered pairs (x, y) satisfying either $x = c$ or $y = mx + b$.
3. The plane is all possible ordered pairs (x, y) of real numbers.
4. The distance $PQ = \sqrt{(x_2 - x_1)^2 + (y_2 - y_1)^2}$ where P is (x_1, y_1) and Q is (x_2, y_2).
5. The midpoint of segment PQ is M. $M = (\frac{x_1 + x_2}{2}, \frac{y_1 + y_2}{2})$ where P is (x_1, y_1) and Q is (x_2, y_2).
6. Two lines are parallel if and only if they have the same slope and different y-intercepts.
7. Two nonvertical and nonhorizontal lines are perpendicular if and only if their slopes are negative reciprocals.

The coordinate model tools above furnish about all that is needed in the geometry course to provide coordinate arguments. As an example, consider the theorem that the segment joining the midpoints of the two sides of a triangle is parallel to and has half the length of the third side (fig. 7.1).

Teaching Matters: One way to discover theorems like this one is through investigation with drawing and measuring utilities. By constructing several arbitrary triangles and measuring appropriate segments, students can begin to organize the results and make a conjecture on the basis of their observations.

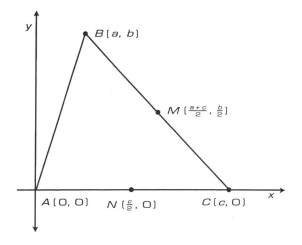

Fig. 7.1

Teaching Matters: *Drawing and measuring utilities are useful computer tools for investigating the properties of figures. But you should discuss their limitations with the students. Suppose, for example, that the utility gave measures of 14.7 and 7.3 for the base and the midline of a triangle. Is this a counterexample to the midpoint connector theorem? No, it simply means that the limits of the software for measuring have been reached. Impress on the students that correct reasoning cannot be negated by limited computer software.*

Coordinate argument:

1. Set up the triangle on the coordinate system.
2. Find the midpoints of two sides—say $\overline{BC}$ and $\overline{AC}$.
3. The midpoint of $\overline{BC}$ is $(\frac{a+c}{2}, \frac{b}{2})$. The midpoint of $\overline{AC}$ is $(\frac{c}{2}, 0)$.
4. Compare the slopes of and :

   For $\overleftrightarrow{AB}$: $\dfrac{b-0}{a-0} = \dfrac{b}{a}$

   For $\overleftrightarrow{MN}$: $\dfrac{\frac{b}{2}-0}{\frac{a+c}{2}-\frac{c}{2}} = \dfrac{b}{a}$

   The segments are parallel.

5. Check the lengths: $AB = \sqrt{a^2 + b^2}$ $MN = \sqrt{[(a+c)/2 - c/2]^2 + (b/2)^2}$
$$= \sqrt{(a^2 + b^2)/4}$$
$$= (\sqrt{a^2 + b^2})/2 = AB/2$$

Synthetic argument:

1. M and N are midpoints of $\overline{BC}$ and $\overline{AC}$, respectively.
2. Thus $2 \cdot CM = CB$, $2 \cdot CN = CA$, and $CB/CM = CA/CN = 2$.
3. $\angle C$ is common to $\triangle ABC$ and $\triangle NMC$. Thus $\triangle ABC \sim \triangle NMC$ by SAS similarity.
4. Thus $\angle MNC = \angle BAC$ and $\overleftrightarrow{MN} \parallel \overleftrightarrow{AB}$ by corresponding angles.
5. Since $\triangle ABC \sim \triangle NMC$, $AB/MN = CB/CM = 2$, or $AB = 2 \cdot MN$.

Try this: *Prove that the altitude from the vertex angle of an isosceles triangle contains the midpoint of the opposite side.*

The following theorems lend themselves to coordinate methods:

A. *The diagonals of a rectangle are equal in length.*
B. *The segment joining the midpoints of sides joining parallel sides of a trapezoid is parallel to the parallel sides.*
C. *The segment in part B is equal in length to one-half the sum of the lengths of the parallel sides.*
D. *The diagonals of a parallelogram bisect each other.*
E. *The segments that join the midpoints of opposite sides of a quadrilateral bisect each other.*
F. *The segments that join the midpoints of the sides of a rhombus form a rectangle.*
H. *In an isosceles triangle, the medians to the congruent sides are congruent.*
I. *The midpoint of the hypotenuse of an isosceles right triangle is equidistant from the legs.*
J. *The medians of a triangle are concurrent.*
K. *The altitudes of a triangle are concurrent.*
L. *The perpendicular bisectors of the sides of a triangle are concurrent.*

Assessment Matters: *Preparing an argument for any of these statements might serve as a test item. Be sure to give partial credit for good reasoning even if the argument is not entirely accurate. Also give the students plenty of time. Argument preparation is a form of nonroutine problem solving, which requires time for thought and planning.*

Here are three illustrative situations that can be modeled using coordinates (Joint Committee 1980, pp. 161, 167, 169):

1. Trees grown for sale as Christmas trees should stand at least five feet from one another while growing. If the trees are grown in parallel rows, what is the smallest allowable distance between rows?

2. The floor plan of a living room of a house is shown in figure 7.2. There are many ways to describe the shape and size of this room. One way is to say that each unit is one foot and to say that consecutive vertices of the octagon are A(0, 0), B(8, 0), C(8, –7), and so on. Complete the coordinate description, if each corner is a right angle. Is there a more convenient location for the origin?

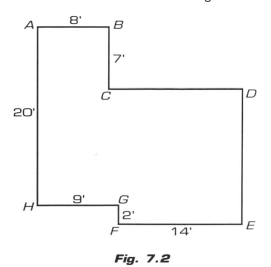

Fig. 7.2

3. Suppose we have paint that is two parts blue and one part white. This mixture is represented by $b = 2w$, where b and w are the amounts of blue and white paint in the mixture. What does the equation $b = 3^w/_2$ represent? Graph these equations with the w-axis horizontal and the b-axis vertical. Consider the points $A(3, 6)$ and $B(1, 1.5)$. How much paint does each of these represent? Which mixture is each? What point corresponds to mixing all of A with all of B? What point corresponds to mixing half of A with half of B? In what region will any point lie that represents a mixture of A and B?

There are many other situations that are easily represented and analyzed using coordinates. As you work through the geometry course, try to identify more. You can almost immediately discard any theorem that involves congruence of angles. The coordinate representation of angle measures is not convenient to use in geometric arguments at this level.

REASONING WITH TRANSFORMATIONS

The major role transformations play in reasoning about shapes is that of symmetry, either bilateral or rotational. If a shape is known to be symmetric, then there is a correspondence of the vertices induced by the symmetry. This correspondence allows us to conclude that pairs of angles are congruent or that pairs of segments are congruent.

Try this: Given a regular hexagon ABCDEF, use symmetry to show that diagonals $\overline{AD}$ and $\overline{BE}$ are congruent.

For example, we know that a rectangle is reflection symmetric about a perpendicular bisector of a side. (See fig. 7.3.) Thus $r(ABCD) = DCBA$. From this it follows that $r(A) = D$ and $r(C) = B$. Thus, $r(\overline{AC}) = \overline{DB}$. We conclude that the diagonals of a rectangle are congruent. For more ideas on how symmetry can be used in the study of shape, consult Senechal (Steen 1990, pp. 150–58).

Another way that transformations may be helpful is in problem solving. For example, consider the following problem:

Given an angle A and a point P interior to angle A, construct a circle containing P that is tangent to the sides of angle A.

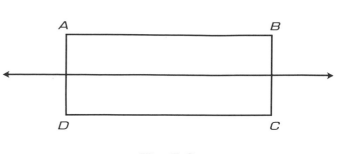

Fig. 7.3

Solution: Consider figure 7.4. We know that the tangent circles have center on the bisector of angle *A*. Construct that bisector. Randomly choose a point on the bisector and construct a circle tangent to the sides of angle *A*. If circle *X* contains *P*, you are done, but that is unlikely.

Draw *AP*, which intersects circle *X* in *P'* and *P"*. Now we have two choices. Map either *P"* or *P'* onto *P* by using a dilation. Say we choose *P'*, then the magnitude of the dilation is *AP/AP'*. Using this transformation, map *X* onto *X'*. Use *X'* as a center and radius *X'P* to draw the desired circle.

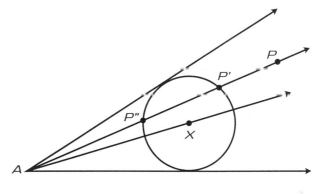

Fig. 7.4

Here are several more problems that can be solved using transformation techniques:

A. In a triangle *ABC*, draw a square so that one side is on $\overline{AB}$ and the other vertices are on $\overline{AC}$ and $\overline{BC}$.

B. Given two parallel lines and a point *A* on one of them. Construct an equilateral triangle with a vertex at A, one side on the parallel containing A, and one vertex on the other parallel.

C. Given a point A and lines *m* and *n*. Construct an equilateral triangle with one vertex at *A* and the others on *m* and *n*.

D. Given *P* and two circles C_1 and C_2. Find points on C_1 and C_2 such that *P* is the midpoint of $\overline{C_1C_2}$. (Hint: Use a half-turn.)

E. Given a line *m* and two circles C_1 and C_2. Construct a square with opposite vertices on *m* and one of the remaining vertices on each of the circles C_1 and C_2.

F. Given two parallel lines and points *A* and *B* between them. Construct the path of a light beam that begins at *A* and illuminates *B* after reflecting twice in each of the parallel lines.

G. A light issues from *A* and reflects off lines *m* and *n* to illuminate *B*. Draw the path followed by the light.

Assessment Matters: Writing a justification of the solution of any of these problems might serve as a test item. The suggestions concerning partial credit and the need for sufficient time, which were given for coordinate problems, apply here as well.

Commonly in geometry, congruence is defined for segments, angles, triangles, and circles. For segments and angles, equal measures imply congruent figures and the converse. Thus equality and congruence are hard to separate. For the circle, congruence is a result of congruent (equal measure) radii. Thus equality of lengths is again central.

For the triangle, the equality of the measure of any *one* quantity is not sufficient to ensure congruence. For example, we know that congruent bases (fig. 8.1a), equal perimeters (fig. 8.1b), equal areas, and congruent angles, by themselves, do not ensure that triangles are congruent. In the case of angles, all corresponding angles may be congruent, and the two triangles are not congruent.

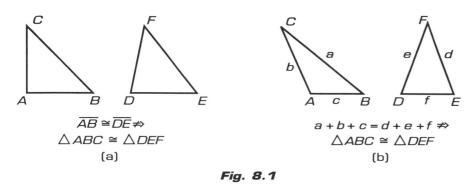

$$\overline{AB} \cong \overline{DE} \not\Rightarrow$$
$$\triangle ABC \cong \triangle DEF$$
(a)

$$a + b + c = d + e + f \not\Rightarrow$$
$$\triangle ABC \cong \triangle DEF$$
(b)

Fig. 8.1

Try this: On graph paper draw a pair of noncongruent triangles with one side in each triangle congruent; with equal perimeters; and with equal areas. Draw two noncongruent triangles in which five parts of one are congruent to five parts of the other.

Since most of the usual figures of geometry may be decomposed into triangles, the centrality of the study of conditions ensuring congruence in triangles is justified. Thus conditions ensuring congruence of triangles such as *SAS* and *ASA* should continue to be proved or assumed and used in original arguments. However, alternative approaches can be used that ensure that transformation ideas become related to congruence of triangles, and, more generally, to congruence of other point sets not traditionally thought of as figures.

One approach is simply an extension of the usual development of congruence. In this development, reflection over a line is defined and the following sequence of definitions and theorems is developed (Jacobs 1987):

1. *Reflection of a pair of points over a line preserves distance.*
2. Definition: *An isometry is a transformation that preserves distance.*
3. *A reflection is an isometry.*
4. *An isometry preserves collinearity and betweenness.*
5. *An isometry preserves angle measure.*
6. *A triangle and its image under an isometry are congruent.*
7. Definition: *A transformation is a translation iff it is the composite of two successive reflections over parallel lines.*
8. Definition: *A transformation is a rotation iff it is the composite of two successive reflections over intersecting lines.*
9. Definition: *Two figures are congruent iff there is an isometry mapping one figure onto the other.*
10. Definition: *A figure has reflection symmetry with respect to a line iff it coincides with its reflection image.*

Teaching Matters: Students need to have many informal experiences that involve reasoning and arguing to support their conjectures before they are likely to understand the need for, or the value of, a formal proof.

Try this: Prove that a translation and a rotation are isometries. Draw images to illustrate that theorems 4, 5, and 6 in this sequence hold for translations and rotations.

Try this: Draw two congruent triangles anywhere in the plane. Find the minimum number of reflections needed to map one onto the other.

11. Definition: *A figure has rotational symmetry with respect to a point if it coincides with its rotation image about the point.*

A second approach to relating congruence to transformations is to replace the ordinary development with one based on transformations. In this approach, reflections are first defined and images are drawn and constructed so that students become familiar with reflecting figures in order to get their images. The reflection symmetric figure is introduced as a figure that is its own image; in the language used in science, it has bilateral symmetry. With this beginning, the following sequence of major steps leads to defining congruence in terms of isometries:

1. A segment is symmetric to its perpendicular bisector (and to the line in which it lies).
2. The angle bisector is the symmetry line for an angle.
3. The line containing the bisector of the vertex angle of an isosceles triangle is a symmetry line for the triangle.
4. Definition: *A translation is the composite of two reflections over parallel lines.*
5. Definition: *A rotation is the composite of two reflections over intersecting lines.*
6. Definition: *Figures F and G are congruent figures iff G is the image of F under a reflection or composite of reflections.*
7. For any figures *F*, *G*, and *H*,
   a. *F* is congruent to *F*;
   b. if *F* is congruent to *G*, then *G* is congruent to *F*;
   c. if *F* is congruent to *G* and *G* is congruent to *H*, then *F* is congruent to *H*.
8. Every isometry preserves angle measure, betweenness, collinearity, and distance.
9. Two segments are congruent iff they have the same length.
10. Two angles are congruent iff they have the same measure.
11. If two figures are congruent, then any pair of corresponding parts are congruent.
12. *SSS* congruence theorem
13. *SAS* congruence theorem
14. *ASA* congruence theorem

Assessment Matters. Reasoning and proof constructing are not easy to assess with traditional paper-and-pencil tests. Informally observing students as they work individually or in groups and routinely requiring students to justify their answers are two alternative ways to assess their reasoning ability.

Try this: Given that two triangles have two sides and the included angle congruent, prove that they are congruent by using transformations.

In the development outlined above, reflection symmetric figures are introduced first, and the symmetry of the kite is done, as well as the symmetries of the segment, angle, and isosceles triangle. These symmetries are later used in the proofs of theorems 12, 13, and 14. This sequence was first published in the 1971 text by Coxford and Usiskin entitled *Geometry: A Transformation Approach*. It is also found in the University of Chicago School Mathematics Project *Geometry* by Coxford, Usiskin, and Hirschhorn (1991).

The strengths of this approach are its generality and its intuitiveness. Intuitively, two figures, whatever they may be, are congruent when they fit exactly. Transformations provide the mathematics to model moving figures around to make them fit exactly. Its generality comes from the fact that the definition of congruent figures does not specify that the figures must be of any special kind. Thus the figure may be a set of discrete points, a parabola, a complex modern art drawing, or even a triangle. This allows us to *apply* the idea of congruence to triangles rather than to define it in terms of triangles.

The approach that is used to introduce transformations is less important than the fact that they are introduced. Transformations permit the work with coordinates to be much richer.

In the coordinate plane, an *n*-gon can be represented with a 2 × *n* matrix in which each column contains the *x*- and *y*-coordinates of a vertex. Thus pentagon *ABCDE* (fig. 8.2) is represented by $\begin{bmatrix} -4 & -2 & 4 & 6 & 0 \\ 0 & 4 & 5 & -1 & -3 \end{bmatrix}$.

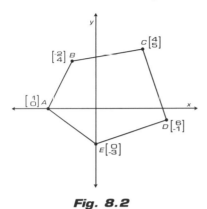

Fig. 8.2

In a similar manner, a solid such as a tetrahedron could be represented by a 3 × 4 matrix in which each column contains the coordinates of a vertex.

In the plane, multiplying a point (a column vector) by a 2 × 2 matrix maps that point onto a point. When both (1, 0) and (0, 1) map onto distinct noncollinear points, the plane maps onto itself in a one-to-one fashion. In this case, the 2 × 2 matrix represents a transformation of the plane onto itself.

Try this: **Consider the matrix** $\begin{bmatrix} 1 & 2 \\ 2 & 4 \end{bmatrix}$. **If (x, y) is any point in the plane, what is the locus of all images of (x, y)?**

The isometries that leave the origin fixed (rotations about the origin, reflections over lines *y* = *mx*, and *x* = 0) have simple 2 × 2 matrix representations. The most common ones are listed in the chart in figure 8.3.

Matrix Representations

reflection over the *x*-axis	$\begin{bmatrix} 1 & 0 \\ 0 & -1 \end{bmatrix}$	$(x, y) \rightarrow (x, -y)$
reflection over the *y*-axis	$\begin{bmatrix} -1 & 0 \\ 0 & 1 \end{bmatrix}$	$(x, y) \rightarrow (-x, y)$
reflection over *y* = *x*	$\begin{bmatrix} 0 & 1 \\ 1 & 0 \end{bmatrix}$	$(x, y) \rightarrow (y, x)$
reflection over *y* = –*x*	$\begin{bmatrix} 0 & -1 \\ -1 & 0 \end{bmatrix}$	$(x, y) \rightarrow (-y, -x)$
rotation of 90° counterclockwise	$\begin{bmatrix} 0 & -1 \\ 1 & 0 \end{bmatrix}$	$(x, y) \rightarrow (-y, x)$
rotation of 180° counterclockwise	$\begin{bmatrix} -1 & 0 \\ 0 & -1 \end{bmatrix}$	$(x, y) \rightarrow (-x, -y)$
rotation of 270° counterclockwise	$\begin{bmatrix} 0 & 1 \\ -1 & 0 \end{bmatrix}$	$(x, y) \rightarrow (y, -x)$

Try this: **Ask students to find each of the matrix representations given in figure 8.3 and to use them to transform the pentagon in figure 8.2.**

Fig. 8.3

Each of these matrix representations can be derived by beginning with a matrix $\begin{bmatrix} a & c \\ b & d \end{bmatrix}$ and two preimage-image pairs. For example, suppose we wish to determine the matrix for a reflection over the line $y = x$. Choose two points, say (2, 3) and (–1, 2), and their images, (3, 2) and (2, –1).

Multiply (2, 3) and (–1, 2) by $\begin{bmatrix} a & c \\ b & d \end{bmatrix}$.

$$\begin{bmatrix} a & c \\ b & d \end{bmatrix}\begin{bmatrix} 2 \\ 3 \end{bmatrix} = \begin{bmatrix} 2a + 3c \\ 2b + 3d \end{bmatrix} = \begin{bmatrix} 3 \\ 2 \end{bmatrix} \text{ and } \begin{bmatrix} a & c \\ b & d \end{bmatrix}\begin{bmatrix} -1 \\ 2 \end{bmatrix} = \begin{bmatrix} -a + 2c \\ -b + 2d \end{bmatrix} = \begin{bmatrix} 2 \\ -1 \end{bmatrix}$$

These provide four equations in four variables. In pairs we have

$$2a + 3c = 3 \qquad\qquad 2b + 3d = 2$$
$$-a + 2c = 2 \qquad\qquad -b + 2d = -1.$$

Solving, we find $a = 0$, $c = 1$, $b = 1$, and $d = 0$. The matrix is shown in figure 8.3.

A simpler way is to use (1, 0) and (0, 1) as the preimage points. Then the values for the columns (a, b) and (c, d) are quickly found to be (0, 1) and (1, 0). Notice that the geometric idea of transformation is connected to the algebraic idea of a matrix, which, in turn, is connected to the solution of systems of linear equations, a fundamental topic in algebra. Thus approaching geometry from multiple perspectives permits stressing connections among mathematical topics as recommended by the NCTM *Curriculum and Evaluation Standards for School Mathematics*.

Matrix representations may also be used to reinforce the relationship between rotations and reflections. Composition of transformations is represented in matrix terms by multiplication of matrices. For example, if we reflect over the x-axis and then the line $y = x$, we should get a rotation of 90° counterclockwise, since the angle from the x-axis to the line $y = x$ is 45°.

$$\begin{bmatrix} 0 & 1 \\ 1 & 0 \end{bmatrix} \cdot \begin{bmatrix} 1 & 0 \\ 0 & -1 \end{bmatrix} = \begin{bmatrix} 0 & -1 \\ 1 & 0 \end{bmatrix}$$

$r_{y=x} \cdot r_{x\text{-axis}}$ = rotation of 90°

Try this: Using your knowledge of sines and cosines of special angles, find matrix representations for counterclockwise rotations of 30° and of 60° about the origin.

If students are familiar with the sine and cosine functions, then the matrix for the rotation through any angle θ can be derived. Its

rotation of θ about the origin is $\begin{bmatrix} \cos\theta & -\sin\theta \\ \sin\theta & \cos\theta \end{bmatrix}$.

For an introductory discussion of computer graphics, transformations, and matrix operations, consult *Matrices* (North Carolina School of Science and Mathematics 1988) in the New Topics for Secondary School Mathematics series.

CHAPTER 9
SIMILARITY FROM MULTIPLE PERSPECTIVES

From the perspective of usefulness, similarity is an important geometric concept. All situations in which scaling is involved depend on similarity. Similarity is the root of trigonometry. Similarity explains why area quadruples when linear dimensions double and why volume increases eight-fold with a doubling of all linear dimensions. Similarity explains why an elephant's legs are so thick and why giants cannot exist. The video "On Size and Shape: Scale and Form," number 17 in the series produced by the Annenberg/CPB project to accompany the text *For All Practical Purposes* (Steen 1988), nicely illustrates and explains such problems of scale. The video discusses the giants of motion pictures and also discusses some of the latest research on dinosaurs and the problems of heat exchange in large animals. Heat exchange is a problem because the production of heat is proportional to the volume of the animal, but the phenomenon of heat loss, which occurs through the skin, is dependent on the surface area.

In previous chapters we began with shapes and moved to coordinate and transformation representations. In this chapter we shall reverse the order; we shall begin with coordinate and matrix representations rather than the more familiar shape-Euclidean approach. Coordinate methods provide an especially convenient vehicle for introducing similarity and the related transformations. This is a viable approach to use in class.

Consider a triangle in the plane with vertices $A(a, b)$, $B(c, d)$, and $C(e, f)$. (See fig. 9.1.) Now *double* each coordinate to get $A'(2a, 2b)$, $B'(2c, 2d)$, and $C'(2e, 2f)$. Draw the image triangle: $\triangle A'B'C'$. Now investigate the following questions:

1. How are the preimage and image segments related?
2. How are the preimage and image distances related?
3. How are the preimage and image angles related?
4. How are the preimage and image triangles related?

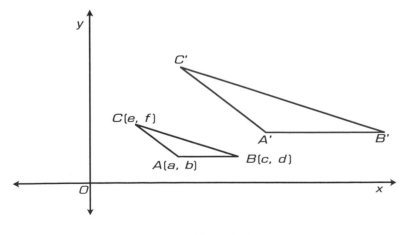

Fig. 9.1

Try this: On graph paper draw a triangle with vertices A(–2, 1), B(4, 2), and C(1, –1). Find the image triangle obtained when each coordinate of each vertex is multiplied by

1. 2.5   2. –2   3. 1/2

4. 1   5. –1.

Compare the area of each image triangle with the area of the original.

The answer to question 2 is that the image distances are double (two times) the corresponding preimage distances. The multiplicative relationship between preimage and image distances should be taken as fundamental, and the language of "k times" should be used instead of the

◆　　　◆　　　◆　　　◆　　　◆　　　◆　　　◆　　　◆

language of ratios. Thus similar figures are those for which correspond-ing preimage and image distances are a given multiple of each other. That multiple is the "constant of similarity," which is often called the ratio of similitude or the scale factor. The software package Geometric Con-nectors: Transformations has the capability to draw similar figures and to read the coordinates of labeled points. It could be used effectively in the study of similarity described here.

Note that the matrix $\begin{bmatrix} k & 0 \\ 0 & k \end{bmatrix}$, when applied to each point in a figure,

effectively multiplies each coordinate by k. This matrix is a representation of the transformation discussed above, the constant of similarity being k.

What are the properties of the transformation illustrated above? Here coordinates are central to the development. First, note that the images of $A(a, b)$ and $B(c, d)$ are $A'(ka, kb)$ and $B'(kc, kd)$. The slope of line AB is $(b - d)/(a - c)$ when $a \ne c$. Also the slope of line $A'B'$ is $(kb - kd)/(ka - kc) = k/k \cdot (b - d)/(a - c)$. Since the slopes of lines AB and $A'B'$ are equal, these lines are parallel. We say that the image of a line is parallel to the preimage line.

Now look at angles. Since an angle is determined by two intersecting lines, and the images of lines are parallel to the preimage lines, an angle and its image are congruent. Notice also that $A(a, b)$ and $A'(ka, kb)$ are on the line through the origin with slope b/a, $a \ne 0$. Thus the origin, A, and A' are collinear. The origin is the center for the transformation. It is its own image.

What about distances? We can easily show that $A'B' = k \cdot AB$. Note also that $OA = \sqrt{a^2 + b^2}$ and that $OA' = \sqrt{k^2 a^2 + k^2 a^2} = k \cdot OA$. This last property is used to define the transformation in synthetic terms as was done on page 8.

Thus we see that the dilation can be developed from a coordinate point of view (with or without the use of the matrix representation). This would be a good application of coordinate methods in the present geometry course, and it would emphasize multiple perspectives and the connec-tions between algebra and geometry.

The dilation, along with the isometries, provides a definition of similar figures that uses the intuitions of motion and expansion.

Definition: *Two figures, F and G, are similar iff F is the image of G under a dilation, an isometry, or a composite of isometries and dilations.*

The usual similarity theorems (postulates in some texts) are easy conse-quences of the transformation definition of similar figures. Below we prove the *SAS* similarity theorem.

Given: $\triangle ABC$ and $\triangle XYZ$; $\angle B \cong \angle Y$, $AB = kXY$, $BC = kYZ$ as in figure 9.2

Prove: $\triangle ABC \sim \triangle XYZ$

Choose O to be any point in the plane. Let S be a dilation with center O and magnitude k. Then $S(\triangle XYZ) = \triangle X'Y'Z'$. Since S preserves angle measure, $\angle Y \cong \angle Y'$. Also, since S multiplies distance by k, $X'Y' = kXY = AB$, and $Y'Z' = kYZ = BC$. It follows that $\triangle X'Y'Z' \cong \triangle ABC$ by *SAS* con-gruence. Thus, $\triangle X'Y'Z'$ is mapped onto $\triangle ABC$ by some isometry T. Finally, $\triangle XYZ$ is mapped onto $\triangle ABC$ by $T \circ S$, a composite of a dilation and an isometry.

Teaching Matters: Students should become completely familiar with the effects of the transformation described here. They should be given extensive practice in plotting preimage and image points and in measuring the corre-sponding distances and the corresponding angles.

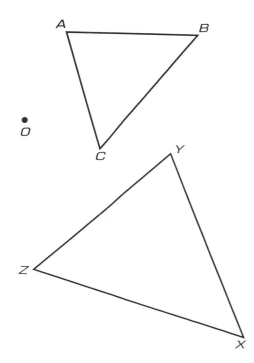

Fig. 9.2

Many people think the square footage of a house with 500 square yards of floor space is $3 \times 500 = 1500$ sq.ft. This erroneous reasoning occurs because the linear relationship of 3 feet a yard is dominant. This illustrates the importance of what we shall call the fundamental theorem of geometry.

If two figures, F and G, are similar with a constant of similarity k, then—

1. the linear dimensions of G are k^1 times those of F;
2. the area of plane regions in G is k^2 times those of F;
3. the volume of space regions in G is k^3 times those of F.

The November 1989 issue of *Student Math Notes*, "Godzilla: Fact or Fiction" (Lott 1989), gives an elementary introduction to this fundamental theorem of geometry.

A recent application of similarity is in the field of fractal geometry. Fractal geometry is used to model mathematically shapes too irregular to be modeled by Euclidean geometry. Some figures that may be easily constructed using computer graphics are *self-similar*. Figures possess self-similarity if and only if portions of the figure are similar to the entire figure. Many fractals are shapes that possess self-similarity. Many of the iterative computer algorithms that produce fractals are simple scaling procedures. For example, consider the equilateral triangle in figure 9.3. We may construct a fractal by repeatedly applying the following procedure to each segment in each figure produced.

Procedure:

1. Trisect each segment of the figure.
2. Construct an equilateral triangle on the middle third.
3. Delete the middle third.
4. Repeat steps 1 through 3.

We can describe this with function notation as $f(\triangle) = $ ★. What we are interested in is $f(f(\triangle))$, then $f(f(f(\triangle)))$, and in general, $f^n(\triangle)$.

Notice that at the end of one iteration, the new equilateral triangles have sides one-third the length of the original and they are similar to the original. The area of each new triangle is one-ninth that of the original, and the perimeter of the new figure is four-thirds the side of the original triangle. Similarity and the constant of similarity are essential in understanding the changes in area and perimeter of fractals constructed as described here.

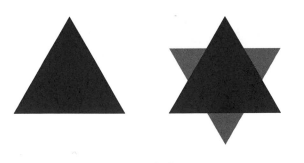

Fig. 9.3

Three classroom-ready activities for introducing students to the exciting world of fractal geometry follow.

*Try this: **Prove the AA similarity theorem by using the transformation definition.***

*Try this: **Suppose two humans, Al and Bob, are similar and the constant of similarity for mapping Al onto Bob is 7/6. You are given some information about Al or Bob. You are to determine the corresponding information about the other.***

1. *Al is 6 feet tall. How tall is Bob?*
2. *Bob has a 36-inch waist. What does Al's waist measure?*
3. *Al weighs 150 pounds. What does Bob weigh?*
4. *It takes 2 yards of cloth to make Al a sport coat. How much cloth will it take to make Bob one?*
5. *Bob eats 3 pounds of food a day. How much will Al eat?*
6. *Bob wears a size 12 shoe. What size does Al wear?*

*Try this: **Duplicate the beginning of the fractal construction in figure 9.3 and advance it for four iterations. What is the sequence of the perimeters and of the areas of the shapes?***

A fractal can be generated by a pattern of iteration. This fractal design is called the Sierpinski carpet after the mathematician who invented it in 1916. The general rule is to start with a square and take a square out. Look at the first iteration and describe the rule that was used to determine the size of the square that was removed. Now compare the first two iterations and describe the rule that was used to construct the second from the first. Apply the rule you have stated to construct the third iteration in the space provided.

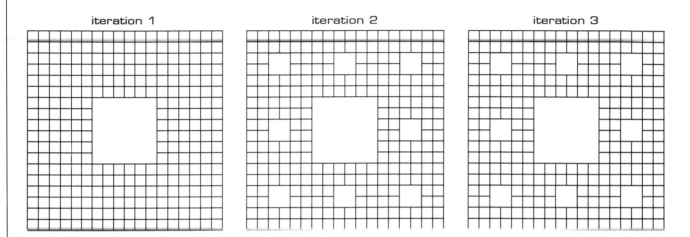

iteration 1   iteration 2   iteration 3

Now examine the third iteration you have constructed, and record the length of the side of the new squares you drew. Compare this length to the lengths of the sides of the previous squares. Write the lengths of the sides of all the squares in descending order. If you construct the fourth iteration, what will the lengths of the sides of the squares need to be? Now look at the first iteration again. What is the area of the square that was removed? What is the area of each individual square that was removed in the next two iterations? Write these areas in descending order. What is the area of each individual square to be removed in the fourth iteration?

Challenge: Find the perimeter of all the squares in the third iteration. Find the area of the figure that remains once all the squares are removed in the third iteration.

You can make some other fractals by trying other patterns. Here is one more example.

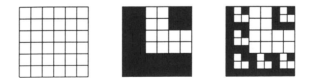

Can you make one using other plane figures? What types of scale factors did you use?

ACTIVITY 17
FRACTAL CURVE

This fractal design will generate a Peano curve. The first iteration is given on the top left of figure 1. To obtain the second iteration, similar "hooks" with sides $\frac{\sqrt{2}}{2}$ are placed between the two pairs of dots as shown in the bottom left diagram in figure 1. The third iteration is made up of four similar hooks that are placed similarly in iteration 2.

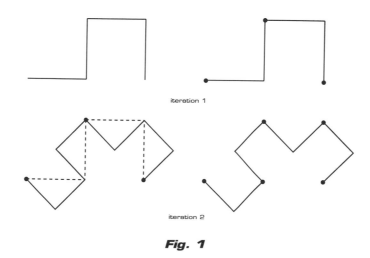

iteration 1

iteration 2

Fig. 1

If the length of each segment that makes up the first iteration is 1, what is the perimeter of the first iteration? What is the perimeter of the second iteration? Draw iteration 3. What are the lengths of these segments? The perimeter? Write the lengths of the segments used in the three iterations in descending order. Predict the length of the segments used in the tenth iteration and the nth iteration.

Figure 2 is a variation, designed by Giamati, on the fractal above. To complete it, a shaded triangle should also be drawn between the dots. The second iteration is constructed much like the second iteration in the design above. Draw the second iteration.

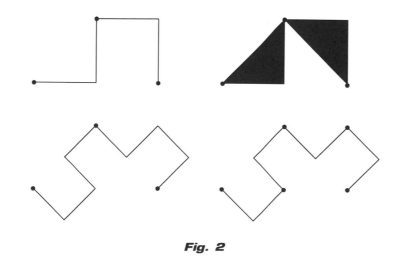

Fig. 2

What is the area of the shaded region in the first iteration? What is the area of the shaded region in the second iteration? What is the area in the third iteration? Are the areas increasing or decreasing? Predict the area in the fifth iteration and in the nth iteration.

ACTIVITY 18
THE SIERPINSKI TRIANGLE REVISITED

The Sierpinski triangle can be created by using dilations and isometries as well as by using the computer program in the chaos game. You may begin with an arbitrary triangle. An equilateral triangle is used for the procedures described below.

1. Draw an equilateral triangle.
2. Reduce the triangle by a factor of ½. Make three copies of the reduced triangle.
3. Place the three reduced similar triangles on the original, one at each vertex.
4. Eliminate the remaining portion of the original triangle by blackening it.

Your work should result in the figure below.

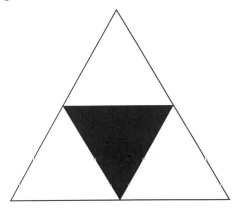

Answer the following questions:

1. Let the area of the original triangle be 1. What area remains? What area has been removed?
2. Let the side of the original triangle be 1. What is the perimeter of the figure with the dark region removed?

Repeat steps 1 through 4 of the original procedure for each of the triangular regions remaining in the figure above. Sketch the result of your work.

Answer the following questions:

1. What is the area of the remaining triangular region?
2. What is the perimeter of the new "holey" triangular region?
3. What would the next iteration of the procedure look like? Make a sketch.
4. Write an expression for the area of the Sierpinski triangle after carrying out the procedure n times.
5. Write an expression for the perimeter of the Sierpinski triangle after carrying out the procedure n times.
6. How would your expressions differ if you began with a triangle other than an equilateral triangle?
7. Compare the figure you are creating with the figure generated by the chaos game.

CHAPTER 10
REASONING, JUSTIFICATION, AND PROOF

Throughout this volume, reasoning, justification, and proof in geometric situations have been illustrated many times. As used here, reasoning is the process of thinking about a mathematical question; a justification is a rationale or argument for some mathematical proposition; and a proof is a justification that is logically valid and based on initial assumptions, definitions, and proved results. A justification may be less formal than a proof. It may consist of a set of examples that seem to support the proposition, or it may be an intuitive argument. These three concepts are related in that we use reasoning to seek a justification of a proposition, which we turn into a proof if we can.

The *Curriculum and Evaluation Standards for School Mathematics* (NCTM 1989) identifies geometry as a continuing important area in the mathematical sciences appropriate for the school mathematics curriculum. Geometry contributes to the development of broadly useful skills in visualization, pictorial representation, and application. It also is a prime area in which mathematical reasoning, justification, and proof may be taught, understood, and practiced. The *Curriculum and Evaluation Standards* suggests that mathematical reasoning be applied in synthetic, in coordinate, and in transformation contexts. The previous pages of this work have illustrated reasoning, justification, and proof as they apply to these areas.

The *Curriculum and Evaluation Standards* suggests further that deduction in geometry should move toward shorter sequences of theorems that are based on sets of "local axioms" that will head the shorter deductive sequences. The full implementation of this suggestion will need to await further work, but within the present geometry curriculum, we can organize some topics so that they have the "flavor" of being organized by "local axiomatics." The *Curriculum and Evaluation Standards* suggests one illustration that begins with the assumption of the *SSS*, *SAS*, and *ASA* triangle congruence statements and the "parallel lines imply alternate interior angles congruent" assertion. Given the definition of a parallelogram as a quadrilateral with two pairs of opposite sides congruent, the following theorems can be proved:

1. *Opposite sides of a parallelogram are congruent.*
2. *Opposite angles of a parallelogram are congruent.*
3. *Diagonals of a parallelogram bisect each other.*
4. *A diagonal of a parallelogram forms two congruent triangles.*
5. *The diagonals of a rhombus are perpendicular.*

Several more theorems about the special types of parallelograms can be proved also. In particular these theorems follow:

6. *Opposite sides of a rectangle are congruent.*
7. *Opposite sides of a rhombus are congruent.*
8. *Diagonals of a rectangle bisect each other.*
9. *Diagonals of a rhombus bisect each other.*
10. *Diagonals of a rectangle are congruent.*
11. *Diagonals of a square bisect each other.*
12. *Diagonals of a square are congruent and perpendicular.*

These last seven theorems follow directly from the assumptions made.

Try this: The maximum number of nonoverlapping regions formed when four points on a circle are connected is 8, or $2^n - 1$ when n = 4. Verify this. Draw new circles, and try values of n that include 1, 2, 3, and 5. Does the pattern always hold? Do the examples constitute a justification? A proof?

They may also be deduced immediately from the first five theorems above if the quadrilateral hierarchy presented in chapter 5 is assumed or derived. This is true because each of theorems 6 through 12 is a theorem about a quadrilateral lower in the hierarchy than the statements made in theorems 1 through 5.

An approach to local axiomatics that makes extensive use of coordinates may be found in the work of Craine (1985). Four statements are assumed:

a. The x-axis and the y-axis are perpendicular.
b. The Pythagorean theorem and its converse are true.
c. Every line has an equation of the form $ax + by = c$ (provided both a and b are not 0) and the converse.
d. Any polygon may be placed with one vertex at the origin and one side on the positive x-axis.

Craine continues to show how six fundamental theorems may be proved from these assumptions. Finally, a set of fourteen theorems and five lemmas are proved. This approach is worthy of examination, but it has not been included in a published textbook for general use.

Another way to implement the idea of local systems is to begin with a definition of a figure, and then to require the students, working in small groups (perhaps cooperatively organized), to develop a list of conjectured properties of the figure, to justify each conjecture informally or with formal proof, to list the assumptions needed in order to justify each conjectured property, and to present an organized list of assumptions, additional definitions, sequence of theorems, and justifications as the final product. For example, the properties of a parallelogram lend themselves to this approach. First, define a parallelogram:

> A parallelogram is a quadrilateral with two pairs of opposite sides parallel.

Then ask the students to investigate the parallelogram in an attempt to list all possible properties of the parallelogram. In the instructions, make sure that you indicate that students may measure segments and angles and may draw any extra (auxiliary) lines that they wish to help them in their search for properties. Once they have found some conjectured properties, they are to try to prove or to refute each using other geometric truths previously developed in the course. Once each group has its theorem sequence and assumptions, a common theorem sequence and assumption list should be worked out by the students. The students should "discover" and "justify" theorems 1 through 4 on page 61, as well as the following new theorem:

> The two diagonals of a parallelogram divide the figure
> into two sets of congruent triangles.

If we also permit the investigation to include special quadrilaterals, the remaining theorems above could also result.

A nice change of pace on this activity is to give half the class the definition of a parallelogram given above and to give the other half the following definition:

> A parallelogram is a quadrilateral with one pair
> of congruent and parallel opposite sides.

Each group works on the project with similar end objectives. However, when the activity is completed, there will be two sets of theorem sequences and two sets of assumptions needed for the development of the theorems. These should be discussed with the class as a whole to point out differences and similarities. The students will notice that the definition for one group will be a theorem for the other group. This allows you to point out that the choice of definition is one that the mathematician needs to make and that different choices lead to different sequences and assumption needs.

The activity described above for parallelograms may be successfully replicated for properties of circles, their chords, and their radii. Here, however, alternative definitions for the circle are not available. The results of this investigation should include some or all of the following:

1. *A chord and the center of a circle determine an isosceles triangle.*
2. *The perpendicular to a chord from the center contains the midpoint of the chord.*
3. *The perpendicular bisector of a chord contains the center of the circle.*
4. *The segment joining the center to a midpoint of a chord (not a diameter) is perpendicular to the chord.*
5. *The center of a circle is the intersection of the perpendicular bisectors of two chords of the circle.*
6. *Circles with the same radius (diameter) are congruent.*

If we also add the definition of a central angle and the degree measure of an arc, then additional possible investigations and results present themselves.

Other topics that may be organized as illustrated above include the following:

1. What are the relationships among the angles formed by parallel lines?
2. What are the sufficient conditions (other that those of the definition) for a quadrilateral to be a rhombus, a parallelogram, a rectangle, a kite, or a square?
3. What are the conditions on two triangles that ensure that they are congruent? Are similar?
4. What special features do right triangles have?
5. What is a general formula for calculating the area of any regular polygon?
6. What special properties do angle bisectors of interior and exterior angles of triangles have, if any?
7. If three lines containing the vertices of a triangle meet in a point, what can be said about the segments into which these lines divide the opposite sides of the triangle?

The examples above are provided to expand on the idea presented in the *Curriculum and Evaluation Standards* that there should be more attention paid to local axiomatics. In the geometry course of today, this can best be approximated by looking in some detail at clusters of theorems and their associated related theory. In addition to the value of the student mathematical activity inherent in the examples above, there is the further value that the students learn that some topics need only a small number of key assumptions for their development. Thus, even though geometry is founded on a large axiom system, parts of it use only small parts of the large system.

Try this: Given that a parallelogram is a quadrilateral with one pair of congruent opposite sides that are also parallel, and given that a diagonal of a parallelogram is a segment determined by nonadjacent vertices,

1. *find as many properties of a parallelogram and its diagonals as you can;*
2. *prove each conjectured property that you can noting what other geometric propositions are needed to complete the proof;*
3. *organize your theorems into a logical sequence in which theorems about parallelograms already proved are used to prove later theorems in your sequence.*

Teaching Matters: The drawing and measuring utilities discussed earlier are useful tools in any investigation of the properties of a figure.

FINAL COMMENTS

In this volume of the Addenda series, we have tried to illustrate how the perspectives of synthetic, coordinate, and transformation geometry can be woven into the present geometry course. We have done this by discussing procedures that can be used with standard geometric topics that all of us teach. We have also suggested that we need to offer students more chances to investigate and think about geometric questions on their own or in small groups. To this end, we have included a number of activity sheets throughout the book. Not all these sheets describe activities that can be completed in a single day, because we think that longer investigations of significant mathematics are appropriate for our youth. Additionally, some of these investigation sheets call for materials not available in all schools. We think that such materials should be in the well-equipped geometry classroom. Some materials considered essential to teach contemporary geometry are listed below so that you can begin now to lobby your department chair and administration to equip your classroom properly:

Graph paper

Ruler

Compass

Protractor

Computer terminal and projection device

Computer software: drawing and measuring utilities

Graphing calculators with matrix capabilities for all students

Miras for use with transformation activities

Models of all solids

Geoboards or dot paper to simulate geoboards

Templates of geometric figures

Templates of figures that tessellate

REFERENCES

Joint Committee of the Mathematical Association of America and the National Council of Teachers of Mathematics. *A Sourcebook of Applications of School Mathematics.* Reston, Va.: The Council, 1980.

Coxford, Arthur, and Zalman Usiskin. *Geometry: A Transformational Approach.* River Forest, Ill.: Laidlaw Brothers, Publishers, 1971.

Coxford, Arthur, Zalman Usiskin, and Daniel Hirschhorn. *University of Chicago School Mathematics Project: Geometry.* Glenview, Ill.: Scott, Foresman & Co., 1991.

Craine, Timothy V. "Integrating Geometry into the Secondary Mathematics Curriculum." In *The Secondary School Mathematics Curriculum,* 1985 Yearbook of the National Council of Teachers of Mathematics, edited by Christian R. Hirsch, pp. 122–27. Reston, Va.: The Council, 1985.

Crowe, Donald. *Symmetry, Rigid Motions and Patterns: HiMAP Module 4.* Arlington, Mass.: Consortium for Mathematics and Its Applications (COMAP), 1986.

Crowe, Donald W., and Thomas M. Thompson. "Some Modern Uses of Geometry." In *Learning and Teaching Geometry, K–12,* 1987 Yearbook of the National Council of Teachers of Mathematics, edited by Mary M. Lindquist, pp. 101–12. Reston, Va.: The Council, 1987.

Hirsch, Christian R., Harold L. Schoen, Andrew J. Samide, Dwight Coblentz, and Mary Ann Norton. *Geometry.* Glenview, Ill.: Scott, Foresman & Co., 1990.

Jacobs, Harold R. *Geometry.* New York: W. H. Freeman & Co., Publishers, 1987.

Kenney, Margaret. "Logo Adds a New Dimension to Geometry Programs at the Secondary Level." In *Learning and Teaching Geometry, K–12,* 1987 Yearbook of the National Council of Teachers of Mathematics, edited by Mary M. Lindquist, pp. 85–100. Reston, Va.: The Council, 1987.

Lott, Johnny W., ed. "Godzilla: Fact or Fiction?" *NCTM Student Math Notes,* November 1989.

Martin, George E. *Transformation Geometry: An Introduction to Symmetry.* New York: Springer-Verlag, 1982.

National Council of Teachers of Mathematics. *Curriculum and Evaluation Standards for School Mathematics.* Reston, Va.: The Council, 1989.

North Carolina School of Science and Mathematics. *Matrices.* New Topics for Secondary School Mathematics series. Reston, Va.: National Council of Teachers of Mathematics, 1988.

Pohl, Victoria. "Visualizing Three Dimensions by Constructing Polyhedra." In *Learning and Teaching Geometry, K–12,* 1987 Yearbook of the National Council of Teachers of Mathematics, edited by Mary M. Lindquist, pp. 144–54. Reston, Va.: The Council, 1987.

Ranucci, Ernest R., and Joseph L. Teeters. *Creating Escher-Type Drawings.* Palo Alto, Calif.: Creative Publications, 1977.

Senechal, Marjorie. "Shape." In *On the Shoulders of Giants: New Approaches to Numeracy,* edited by Lynn A. Steen, pp. 139–81. Washington, D.C., National Academy Press, 1990.

Steen, Lynn A., ed. *For All Practical Purposes: Introduction to Contemporary Mathematics.* New York: W. H. Freeman & Co., Publishers, 1988.

Teeters, Joseph L. "How to Draw Tessellations of the Escher Type." *Mathematics Teacher* 67 (April 1974): 307–10.

Appendix

SOLUTIONS AND HINTS

Activity 1: Midsegment of a Triangle

Data will vary, but a major final conjecture should be that the line segment joining the midpoints of two sides of any triangle is parallel to the third side and half its length. The proof can be found in any school geometry book.

Activity 2: Line Parallel to a Side of a Triangle

Data collected will vary depending on where the labeled point is. Possible conjectures: The parallel segment divides the two sides proportionally. The ratio of the parallel segments is the same as the ratio of the sides of the two triangles.

Activity 3: Reflections over the Line $y = x$ on the Coordinate Plane

1. The labeling and sketching will vary. The relationship between a point and its image over $y = x$ is that (x, y) maps onto (y, x) or that the line $y = x$ is the perpendicular bisector of the segment determined by (x, y) and its image.

2. The image point is always (y, x) when the preimage point is (x, y).

3. If $A(x, y)$ is reflected over $y = x$, then its image A' has coordinates (y, x). Justifications will vary. Some may involve careful drawings, others may use the coordinate system and the distance formula.

Activity 4: General Pythagoras

1. The Pythagorean relation exists among the areas.

2. Again the sum of the areas of the equilateral triangles on the legs equals the area of the equilateral triangle on the hypotenuse. Justification uses the fact that the altitudes are proportional to the sides (that is, the legs and hypotenuse of the right triangle). The areas sum as desired because a-squared plus b-squared equals c-squared in $\triangle ABC$, and the height of each equilateral triangle is $\sqrt{3/2}$ times a leg or hypotenuse.

3. Here the height of each triangle equals a, b, or c. Thus the areas sum as desired.

4. Here the radius is half of a, b, or c. Again the areas sum as desired because of the relationship among a-squared, b-squared, and c-squared.

5. Here similarity assures that the height of each triangle is proportional to a leg or hypotenuse. Thus the sum of the areas on the legs equals the area on the hypotenuse because of the Pythagorean relation.

6. The conjecture desired is "If similar figures are constructed on the legs and hypotenuse of a right triangle, then the sum of the areas of the figures on the legs is the area of the figure on the hypotenuse."

Activity 5: Midpoints on a Triangle

1 and 2. The proofs are quite simple and are based on the midpoint connector theorem. For the second one, note that there are three different parallelograms, each of which contains $\triangle DEF$.

3. One way to proceed is to perform a half-turn of triangle 1 about D, the midpoint of side AB. Then translate this new triangle to new positions to complete the larger desired triangle. A second way to proceed is to rotate triangle 1 about the midpoints of each side, the result being the final triangle.

Activity 6: The Chaos Game

The result of the chaos game is an approximation to the Sierpinski triangle. It looks like the diagram at the left.

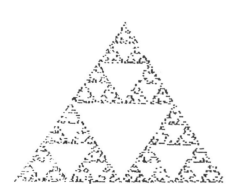

◆   ◆   ◆   ◆   ◆   ◆   ◆   ◆

Activity 7: Midpoints of the Sides of a Quadrilateral

Original figure	New figure
1. parallelogram	parallelogram
2. trapezoid	parallelogram
3. isosceles trapezoid	rhombus
4. kite	rectangle
5. square	square
6. rhombus	rectangle
7. rectangle	rhombus

In general, the figure formed when the midpoints of any quadrilateral are joined is a parallelogram. Special cases may occur when special conditions hold, such as when the diagonals of the original quadrilateral are congruent or perpendicular.

Activity 8: Areas of Parallelograms

1.

	Length of base	Height	Area
a.	2	2	4
b.	3	1	3
c.	3	2	6
d.	$\sqrt{2}$	$2\sqrt{2}$	4
e.	3	3	9
f.	$2\sqrt{2}$	$3\sqrt{2}/2$	6

2. Area = length · height

4. Area = $h \cdot (B + b)/2$, where B and b are bases

Activity 9: Equilic Quadrilaterals

1. One way to draw the equilic quadrilateral is to draw a 60° angle and then choose a length on one side. Choose the same length on the other side and connect the corresponding endpoints. Be sure you do not get the special case of the isosceles trapezoid.

2. Responses will vary.

3. a. Yes, the isosceles trapezoid with 60° angles at the base

   b. No, the equilic quadrilateral would be a parallelogram.

   c. Yes

   d. Yes, the isosceles trapezoid

   e. 240° at A and B, 120° at C and D

   f. Yes, the other angle will be 30°.

4. JKLM forms a rhombus. △JLM and △KLM are equilateral triangles. This is proved by applying the midpoint connector theorem to segments AL and AM. Note that JKLM is a rhombus with five congruent segments, and the only one not having the length of a side is the diagonal JK.

5. △CDP is equilateral. It is proved by showing that △PAD and △PBC are congruent by SAS. Then △CDP is equilateral because it is isosceles with a 60° vertex angle.

Activity 10: Quadrilateral Investigation

2. A variety of properties will be evident for each figure.

3. The one property that is true for all the figures is that the opposite angles of the quadrilateral WXYZ are supplementary. This is hard for students to find, but it is the one common property. Another way to say the same thing is to say that the vertices of WXYZ lie on a circle.

Activity 11: The Cairo Tessellation

3. You see the 3-3-4-3-4 Archimedean tessellation.

Challenge: Each angle is 90 degrees. Justifications will vary.

Activity 12: Archimedean Duals

1. The new pattern is a tessellation. It is made up of kites.

2. The duals of 3-3-4-3-4 and 3-3-3-4-4 are each made up of pentagons. In the former case, the pentagon is the Cairo pentagon, and in the latter it is like home plate in baseball.

Challenge: The five remaining Archimedean tessellations are coded 3-6-3-6, 3-3-3-3-6, 4-8-8, 3-12-12, and 4-6-12.

Activity 13: Frieze Patterns

3. The patterns given are as follows:

a. mg	b. m1	c. m1	d. 1m	e. 1g	f. m1
g. 1m	h. mm	i. 12	j. mm	k. mm	l. 11

Challenge: M2 means vertical reflection and half-turn. It was not included because all "m2" would be included in the "mg" class. (You might note also that all "mm" patterns are also "mg" patterns, so that the classification scheme leaves some ambiguity. For another system, see Martin [1982, pp. 78–87].)

Activity 14: Visualizing Solids

1. No, once you pass through the fifth straw, you must go back through one straw to pick up the sixth straw.

2. Twelve straws are needed for each. The cube cannot be done with the thread going through each straw only once, but the octahedron can be done in this manner.

3. Challenge: Building a tetrahedron inside a tetrahedron: Build a tetrahedron with straws of length 13 cm. This will be the outer tetrahedron. Using a needle and thread, construct the three medians of each face of the tetrahedron. The medians of each triangle should intersect in one point. The inner tetrahedron can be made with edges of length 3.8 cm. The vertices of the inner tetrahedron will be located at the points of intersection of the medians of the outer tetrahedron faces.

Building a tetrahedron inside a cube: Build an octahedron with straws of length 13 cm. Construct the three medians of each face of the octahedron with thread. The inner octahedron is made with edges of length 5.5 cm. Its vertices are the points of intersection of the medians of each outer octahedron face.

Building an octahedron inside a tetrahedron: Build a tetrahedron using straws of length 13 cm. The octahedron can be constructed by connecting the midpoint of each edge of the tetrahedron to the midpoints of the four edges that touch the first edge. The octahedron will have sides of length 6.3 cm.

Building a tetrahedron inside a cube: Build a tetrahedron with sides of length 19 cm. Build tetrahedra on each face of the tetrahedron so that one vertex of each new tetrahedron will be a vertex of the cube. The remaining three vertices of the cube will be the vertices of the original tetrahedron. Straws of length 13.5 cm can be used to complete each new tetrahedron.

Activity 15: Collapsing Cubes

1. Solutions to the topless box:

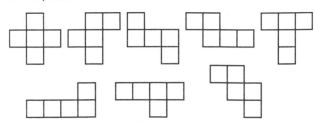

2. These configurations do not fold into a topless cube. The areas are always the same, but the perimeters may differ.

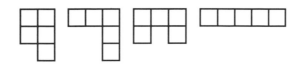

3. Left to the reader

4. Solutions to the cube:

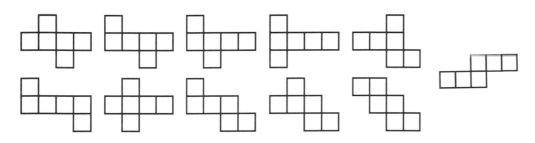

These configurations cannot be folded into a cube:

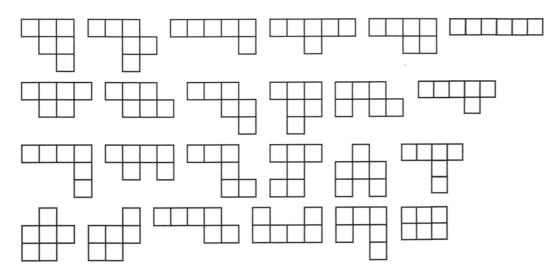

All have the same area. The perimeters differ.

Activity 16: Fractal Carpet

A square of side ⅓ the original was cut out of the center of each square.

The solutions given below assume that the side of the original square is 18.

The length of the side of a square in iteration 3 is ⅔.

The lengths of the sides of the squares in descending order are 18, 6, 2, ⅔.

The length of the side of a square in iteration 4 is ²⁄₉.

The area of the first square removed is 36.

The areas of the next two sized squares to be removed are 4 and ⁴⁄₉.

The area of each individual square removed in iteration 4 is ⁴⁄₈₁.

Challenge: perimeter: $4 \cdot 18 + 4 \cdot 6 + 8(4 \cdot 2) + 8 \cdot 8(4 \cdot ⅔) = 330.67$

area: $324 - (36 + 8 \cdot 4 + 64 \cdot ⁴⁄₉) = 227.56$

iteration 3

Activity 17: Fractal Curve

It is assumed that each segment in the original figure is 1 unit.

Iteration 1 perimeter = 4

Iteration 2 perimeter = $4\sqrt{2}$

Iteration 3 segments are ½ and the perimeter is 8.

In order: 1, $\sqrt{2}/2$, ½

The tenth iteration side is $(\sqrt{2}/2)^9$, the nth iteration is $(\sqrt{2}/2)^n$.

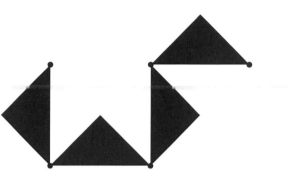

The area of each iterate is 1.

The areas are constant, each being 1.

The third iteration looks like the figure at the left.

Activity 18: The Sierpinski Triangle Revisited

1. ¾, ¼

2. 3 + 3/2 = 4½

Sketches will vary, but each white triangle of the figure shown should have a black triangle removed.

1. $1 - \frac{1}{4} - 3 \cdot \frac{1}{16} = \frac{9}{16}$

2. $3 \cdot 1 + 3 \cdot \frac{1}{2} + 3 \cdot 3 \cdot \frac{1}{4} = 6\frac{3}{4}$

3. Sketches will vary.

4. $1 - 1 \cdot \frac{1}{4} - 3 \cdot \frac{1}{16} - 9 \cdot \frac{1}{64} - \ldots - 3^{(n-1)} \cdot \frac{1}{2}^{2n}$

5. $3 \cdot \left(1 + \frac{1}{2} + \frac{3}{4} + \frac{9}{8} + \frac{27}{16} + \ldots + 3^{(n-2)}/2^{(n-1)} \ldots\right)$, for $n \geq 2$.

6. The expression for the area would be the same. The expression for the perimeter would change, since we would need to let the perimeter of the original triangle be 1. In this case the total perimeter would be $1 + \frac{1}{2} + \frac{3}{4} + \frac{9}{8} + \frac{27}{16} + \ldots$, which is the same as what is in the parentheses in #5.

7. If the chaos game runs long enough, the two figures will be nearly the same, except for a few initial points that initially are outside of the triangle in the chaos game.

BIBLIOGRAPHY

Ball, W. W. R., and H. S. M. Coxeter. *Mathematical Recreations and Essays*. Hong Kong: University of Toronto Press, 1974.

Chapters 4 and 5 of this book deal with unusual problems of a geometric origin, and the ideas related to tessellations are found on pages 105–7. The authors' algebraic approach to the determination of all possible tessellations of one regular polygon is interesting, but they do not treat the problem of tessellations involving more than one polygon.

Bennett, Albert, and Bernard Nelson. *Mathematics: An Informal Approach*. Boston: Allyn & Bacon, 1979.

This book gives a very general and simplified introduction to mappings, and it illustrates the occurrence of geometric concepts in nature. The material presented is easily adaptable to secondary school mathematics in general, and chapter 9 gives a fine account of tessellations.

Bezuszka, Stanley, Margaret Kenney, and Linda Silvey. *Tessellations: The Geometry of Patterns*. Palo Alto, Calif.: Creative Publications, 1977.

Written as a series of activity lessons, this book provides brief explanation and related worksheets in developing key ideas in the study of tessellations. It is geared toward a variety of grade levels.

Bool, F. H., J. L. Kist, J. L. Locher, and F. Wierda. *M. C. Escher: His Life and Complete Graphic Work*. New York: Harry N. Abrams, 1981.

This is the essential reference on the life and art of M. C. Escher! A complete source of biographical information as well as a beautiful collection of all the artist's works, this book can be read and enjoyed from a mathematical or a nonmathematical perspective. The many essays and commentaries enhance the understanding of Escher's growing interest in tessellation and periodic designs. An intriguing book!

Chazen, Daniel, and Richard Houde. *How to Use Conjecturing and Microcomputers to Teach Geometry*. Reston, Va.: National Council of Teachers of Mathematics, 1989.

Practical suggestions for using the Geometric Supposers are given along with suggestions for classroom management.

Coxeter, H. S. M. *Introduction to Geometry*. New York: John Wiley & Sons, 1961.

In chapter 4 of this book, the author presents an excellent summary of two-dimensional crystallography. The discussion is mathematical, clear, and enlightening! The concepts of fundamental regions, lattices, and tessellation-generators are precisely defined and logically motivated. This book is an exceptionally fine source!

Dahlke, Richard, and Robert Fakler. *Applications of High School Mathematics in Geometrical Probability—UMAP Module 660*. Arlington, Mass.: Consortium for Mathematics and Its Applications (COMAP), 1986.

Some probabilistic situations can be easily modeled with geometric models. This work poses probability problems that use geometric figures and their areas or volumes for their solution.

Francis, Richard. *The Mathematician's Coloring Book: HiMAP Module 13*. Arlington, Mass.: Consortium for Mathematics and Its Applications (COMAP), 1989.

The material discusses the question of coloring certain geometric regions, which are made up of nonoverlapping subregions. The ideas are applied to a variety of real situations

Jackiw, Nicholas. The Geometer's Sketchpad. Berkeley, Calif.: Key Curriculum Press, 1991.

Software and reference manual for drawing and measuring figures. Figures may be changed dynamically on the screen and measures are updated and displayed continuously. The most powerful drawing utility to date.

Joint Committee of the Mathematical Association and the NCTM. *A Sourcebook of Applications of School Mathematics*. Reston, Va.: The Council, 1980.

There is a chapter devoted to applications of school geometry as well as chapters on the applications of arithmetic, algebra, trigonometry, and probability.

Kraitchik, Maurice. *Mathematical Recreations*. New York: Dover Publications, 1953.

The most noteworthy aspect of this book with respect to the subject of tessellations is the fine classification of the types of tessellations that exist and the algebraic approach to finding them. The approach is nonformal and provides a good deal of background material for the study of tessellations.

MacGillavry, Caroline H. *Symmetry Aspects of M. C. Escher's Periodic Drawings*. Utrecht: A. Oosthoek's Uitgeversmaatschappij NV, 1965.

The author presents in nontechnical mathematical language the basic concepts of periodic patterns from the viewpoint of symmetry laws and geometric motions. The style is well suited for independent study; step-by-step development of key concepts is the primary approach taken. The author uses a different work of M. C. Escher to illustrate each symmetry group as she explains the principles.

National Council of Teachers of Mathematics. *Geometry in the Mathematics Curriculum*. Thirty-sixth Yearbook of the NCTM, edited by Kenneth B. Henderson. Reston, Va: The Council, 1970.

There are several chapters in this work that will assist the teacher in understanding content presently being suggested for inclusion in school geometry: chapter 5, coordinate approaches; chapter 6, transformation approaches; chapter 8, vector approaches; and chapter 10, an eclectic approach.

_____. *Learning and Teaching Geometry, K–12*. 1987 Yearbook of the National Council of Teachers of Mathematics, edited by Mary Montgomery Lindquist. Reston, Va.: The Council, 1987.

This yearbook discusses many issues related to geometry. It has chapters on the use of Logo and fractals, on strip patterns and group ideas in geometry, and on the use of probability in geometry. This is a good reference for problems and ideas.

Peitgen, Heinz-Otto, Hartmut Jürgens, and Dietmar Saupe. *Fractals for the Classroom*. New York: Springer-Verlag, forthcoming.

This book is especially written for teachers and is intended for the high school and college level. It is to be accompanied by *Fractals for the Classroom: Strategic Activities Volume One* and *Strategic Computer Experiments on Fractals*.

Ranucci, Ernest R., and Joseph L. Teeters. *Creating Escher-Type Drawings*. Palo Alto, Calif.: Creative Publications, 1977.

This book is simply the best sourcebook available for introducing tessellation drawing from a "how to" perspective. It is ideal for use in a variety of classroom settings. There is a logical arrangement of skills and techniques that lead to a deep understanding of the mechanics involved in creating a nonpolygonal tessellation. The explanations included are brief but useful, and the variety of the worksheets makes the actual task of drawing tessellations much easier.

Schwartz, Judah, and Michal Yerushalmy. The Geometric Supposer: Triangles. Newton, Mass.: Education Development Center, 1985.

The software will draw all sorts of triangles, construct lines associated with triangles, and measure distances and angles quite accurately. A key feature is the *redraw* command. This command allows the machine to repeat a construction, which was done by a series of separate commands once, on a new figure.

_____. The Geometric Supposer: Quadrilaterals. Newton, Mass.: Education Development Center, 1985.

Software in the Supposer series for quadrilaterals.

_____. The Geometric Supposer: Circles. Newton, Mass.: Education Development Center, 1986.

Software in the Supposer series for circles.

Steen, Lynn A., ed. *For All Practical Purposes: Introduction to Contemporary Mathematics*. New York: W. H. Freeman & Co., Publishers, 1988.

This text contains three chapters that are especially useful in modernizing the geometry curriculum. Chapter 13 ("Patterns") discusses symmetry, tilings, and patterns. Chapter 14 ("Growth and Form") discusses geometric similarity, scales, and the relation between area and volume. Chapter 21 ("Computer Graphics") provides applications of coordinates in computer science. It also discusses fractals briefly.